The Biology Seagrasses

THE BIOLOGY OF HABITATS SERIES

This attractive series of concise, affordable texts provides an integrated overview of the design, physiology, and ecology of the biota in a given habitat, set in the context of the physical environment. Each book describes practical aspects of working within the habitat, detailing the sorts of studies which are possible. Management and conservation issues are also included. The series is intended for naturalists, students studying biological or environmental science, those beginning independent research, and professional biologists embarking on research in a new habitat.

The Biology of Rocky Shores
Colin Little and F. A. Kitching

The Biology of Polar Habitats
G. E. Fogg

The Biology of Lakes and Ponds
Christer Brönmark and Lars-Anders Hansson

The Biology of Streams and Rivers
Paul S. Giller and Bjorn Malmqvist

The Biology of Mangroves
Peter F. Hogarth

The Biology of Soft Shores and Estuaries
Colin Little

The Biology of the Deep Ocean
Peter Herring

The Biology of Lakes and Ponds, Second ed.
Christer Brönmark and Lars-Anders Hansson

The Biology of Soil
Richard D. Bardgett

The Biology of Freshwater Wetlands
Arnold G. van der Valk

The Biology of Peatlands
Håkan Rydin and John K. Jeglum

The Biology of Mangroves and Seagrasses
Peter Hogarth

The Biology of Mangroves and Seagrasses

Peter Hogarth
Department of Biology
University of York

OXFORD
UNIVERSITY PRESS

OXFORD
UNIVERSITY PRESS

Great Clarendon Street, Oxford OX2 6DP

Oxford University Press is a department of the University of Oxford.
It furthers the University's objective of excellence in research, scholarship,
and education by publishing worldwide in

Oxford New York

Auckland Cape Town Dar es Salaam Hong Kong Karachi
Kuala Lumpur Madrid Melbourne Mexico City Nairobi
New Delhi Shanghai Taipei Toronto

With offices in

Argentina Austria Brazil Chile Czech Republic France Greece
Guatemala Hungary Italy Japan Poland Portugal Singapore
South Korea Switzerland Thailand Turkey Ukraine Vietnam

Oxford is a registered trade mark of Oxford University Press
in the UK and in certain other countries

Published in the United States
by Oxford University Press Inc., New York

© Peter Hogarth 2007

The moral rights of the author have been asserted
Database right Oxford University Press (maker)

First published 2007

All rights reserved. No part of this publication may be reproduced,
stored in a retrieval system, or transmitted, in any form or by any means,
without the prior permission in writing of Oxford University Press,
or as expressly permitted by law, or under terms agreed with the appropriate
reprographics rights organization. Enquiries concerning reproduction
outside the scope of the above should be sent to the Rights Department,
Oxford University Press, at the address above

You must not circulate this book in any other binding or cover
and you must impose the same condition on any acquirer

British Library Cataloguing in Publication Data

Data available

Library of Congress Cataloging in Publication Data

Data available

Typeset by Newgen Imaging Systems (P) Ltd., Chennai, India
Printed in Great Britain
on acid-free paper by
Biddles Ltd., King's Lynn, Norfolk

ISBN 978–0–19–856870–4 (Hbk.)   978–0–19–856871–1 (Pbk.)

10 9 8 7 6 5 4 3 2 1

Preface

Flowering plants dominate the land, providing nutrition, shelter, and stability for a host of organisms, and the basis of all terrestrial ecosystems. Of the hundreds of thousands of species of flowering plants, a mere 100 or so survive in the sea, about equally divided between mangroves and seagrasses. Although not rich in species, both mangroves and seagrasses are, like their terrestrial counterparts, of major ecological importance.

To most people, mangroves call up a picture of a dank and fetid swamp, of strange-shaped trees growing in foul-smelling mud, inhabited mainly by mosquitoes and snakes. Mud, methane, and mosquitoes are certainly features of mangrove forests—as, sometimes, are snakes. They are not sufficient to deter mangrove biologists from investigating an ecosystem of great richness and fascination.

Mangroves are an assortment of tropical and subtropical trees and shrubs which have adapted to the inhospitable zone between sea and land: the typical mangrove habitat is a muddy river estuary. Salt water makes it impossible for other terrestrial plants to thrive here, while the fresh water and the soft substrate are unsuitable for macroalgae, the dominant plants of hard-bottomed marine habitats. The mangrove trees themselves trap sediment brought in by river and tide, and help to consolidate the mud in which they grow. They provide a substrate on which oysters and barnacles can settle, a habitat for insects, and nesting sites for birds. Most of all, through photosynthesis, they supply an energy source for an entire ecosystem comprising many species of organism. Mangroves are among the most productive and biologically diverse ecosystems in the world.

Seagrasses, although not true grasses, generally grow in a grass like way, often locally dominating their environment in what are known as seagrass meadows. They grow intertidally, like mangroves, but also subtidally to depths of tens of metres. Like mangroves, too, seagrasses have adapted to conditions of high salinity and living in soft sediments. They create a habitat, and represent a food source on which many other organisms depend.

With both mangroves and seagrasses I discuss the adaptations to their challenging environment, and the communities of organisms that flourish in and around mangrove forests and seagrass meadows, before turning to more general questions of evolution, biogeography, and biodiversity.

In a final chapter I discuss the relationship between humans and mangroves and seagrasses. Mangroves and seagrasses are of considerable economic benefit. Apart from the direct collection of mangrove products, many commercially harvested species of fish, shrimp, and crab are sustained by mangroves and seagrasses. Unfortunately, the importance of mangroves and seagrasses is not always appreciated, and recent years have seen massive degradation and destruction of both habitats, sometimes deliberate, and in other cases inadvertent. Conservation, restoration, and sustainable management of these important resources are essential. The impact of continuing loss of mangroves and seagrasses seems almost too obvious to need pointing out. Cassandra was fated to predict the future and to have her predictions ignored; biologists sometimes feel they have a similar role.

The productivity and diversity of these remarkable habitats therefore makes them of great interest to biologists and of considerable social and economic value, while degradation and destruction by human activities makes it more than ever essential to understand their significance. Research has advanced considerably in recent years, and the time seems right for an attempt to present our current understanding of the mangrove and seagrass ecology.

My aims in writing this book were two-fold: to share my own enthusiasm for these remarkable ecosystems, and to explain how our understanding is unfolding. Any author depends on the work of others, and I am grateful to numerous colleagues for their help in various ways. In particular, I should like to thank Larry Abele, Liz Ashton, Patricia Berjak, Mike Gee, Rony Huys, Daphne Osborne, Mohammed Tahir Qureshi, and Di Walker. Any errors that remain are, of course, entirely my own.

Writing books has its pleasures, particularly learning about areas of the subject with which one was previously not sufficiently familiar. It also has its disadvantages, and most authors would at some stage agree with the heartfelt—and, in this context, singularly apposite—words of the great American naturalist John James Audubon: 'God . . . save you the trouble of ever publishing books on natural science . . . . I would rather go without a shirt . . . through the whole of the Florida swamps in mosquito time than labor as I have . . . with the pen.'[1] For sustaining me throughout the labours with the pen (and for joining me in the Malaysian swamps in mosquito time) I should especially like to express my gratitude to my wife, to whom this book is dedicated.

P.J.H.
York, September 2006

---

[1] Letter to J. Bachman, 1834, quoted by Ford, A. 1957. *The Bird Biographies of John James Audubon*, pp. vii–viii (Macmillan, New York), and in Rosenzweig (1995).

Contents

Preface   v

1   Mangroves and seagrasses   1

   1.1   Mangroves   2
   1.2   Seagrasses   4

2   Mangrove trees and their environment   8

   2.1   Adaptations to waterlogged soil   8
   2.2   Coping with salt   17
   2.3   The cost of survival   21
   2.4   Inorganic nutrients   24
       2.4.1   Nitrogen   25
       2.4.2   Phosphorus   26
       2.4.3   Nutrient recycling   26
       2.4.4   Does nutrient availability limit growth?   27
   2.5   Reproductive adaptations   29
       2.5.1   Pollination   29
       2.5.2   Propagules   30
       2.5.3   Fecundity and parental investment   33
       2.5.4   Dispersal   34
       2.5.5   Why vivipary?   36
   2.6   Why are mangroves tropical?   38

3   Seagrasses and their environment   39

   3.1   Growth and structure   39
   3.2   Photosynthesis and respiration   41
   3.3   Salinity   42
   3.4   Nutrients   44
   3.5   Reproduction   46
   3.6   Propagule dispersal   47
   3.7   Seagrasses change their environment   48

4 Community structure and dynamics   49

- 4.1 Mangroves: form of the forest   49
  - 4.1.1 Species zonation   52
    - 4.1.1.1 Propagule sorting   54
    - 4.1.1.2 Physical gradients   55
    - 4.1.1.3 Geomorphological change   59
    - 4.1.1.4 Plant succession   59
  - 4.1.2 How different are mangroves from other forests?   62
  - 4.1.3 Do mangroves make mud?   64
- 4.2 Seagrass meadows   67

5 The mangrove community: terrestrial components   71

- 5.1 Mangrove-associated plants   71
- 5.2 Animals from the land   73
  - 5.2.1 Insects   73
    - 5.2.1.1 Insect herbivory   73
    - 5.2.1.2 Termites   77
    - 5.2.1.3 Ants   77
    - 5.2.1.4 Mosquitoes and other biting insects   80
    - 5.2.1.5 Synchronously flashing fireflies   80
    - 5.2.1.6 Other insects   82
  - 5.2.2 Spiders   82
  - 5.2.3 Vertebrates   83
    - 5.2.3.1 Amphibians   83
    - 5.2.3.2 Reptiles   85
    - 5.2.3.3 Birds   89
    - 5.2.3.4 Mammals   93

6 The mangrove community: marine components   98

- 6.1 Algae   98
- 6.2 Fauna of mangrove roots   99
- 6.3 Invertebrates   101
  - 6.3.1 Crustaceans   102
    - 6.3.1.1 Crabs   102
      - Leaf eating by crabs   103
      - Are crabs selective feeders?   107
      - Seedlings   110
      - How important are herbivorous crabs?   112
      - Tree-climbing crabs   113
      - Fiddler crabs   114
      - Reproductive adaptations   118
      - Fiddler patterns in time and space   119
      - The physiology of living in mud   119
      - How stressful is a crab's life?   123

	6.3.1.2 Other mangrove crustacea	124
	6.3.1.3 Crustaceans as ecosystem engineers	125
6.3.2 Molluscs		127
	6.3.2.1 Snails	127
	6.3.2.2 Bivalves	130
6.4 Meiofauna		130
6.5 Fish		133

7 Seagrass communities — 137

7.1 Epiphytes	137
7.2 Molluscs	139
7.3 Crustaceans	140
7.4 Echinoderms	140
7.5 Fish	141
7.6 Turtles	143
7.7 Dugongs and manatees	144
7.8 Birds	145

8 Measuring and modelling — 147

8.1 Mangroves	147
8.1.1 How to measure a tree	148
8.1.2 Biomass	148
8.1.3 Estimating production	151
8.1.4 What happens to mangrove production?	154
8.1.4.1 Microbial breakdown	154
8.1.4.2 Crabs and snails	157
8.1.4.3 Wood	158
8.1.4.4 The role of sediment bacteria	159
8.1.4.5 The fate of organic particles	159
8.1.4.6 Predators	160
8.1.5 Putting the model together	161
8.2 Seagrasses	163

9 Comparisons and connections — 166

9.1 How distinctive are mangrove and seagrass communities?	166
9.2 Mangroves and salt marshes	168
9.3 Interactions	169
9.4 Outwelling	170
9.5 The fate of mangrove exports	172
9.6 Larval dispersal and return	174
9.7 Mangroves, seagrasses, and coral reefs	175
9.8 Commuting and other movement	177

	9.9	Mangroves, seagrasses, and fisheries	178
		9.9.1 Shrimps	178
		9.9.2 Fish	182

10 Biodiversity and biogeography 183

10.1	What, if anything, is biodiversity?	183
10.2	Mangroves	184
	10.2.1 Regional diversity	184
	10.2.2 Origins	188
	10.2.3 Local diversity	192
	10.2.4 Genetic diversity	196
	10.2.5 Diversity of the mangrove fauna	197
10.3	Seagrass biogeography and biodiversity	200
10.4	Diversity and ecosystem function	202

11 Impacts 206

11.1	Mangroves	206
	11.1.1 Uses of mangroves	208
	11.1.1.1 Direct uses	208
	11.1.1.2 Indirect uses	209
	11.1.1.3 Coastal protection	210
	11.1.1.4 Ecotourism	211
	11.1.2 Sustainable management: the case of the Matang	211
	11.1.3 Shrimps versus mangroves?	213
	11.1.4 Mangroves and pollution	216
	11.1.5 Hurricanes and typhoons	220
	11.1.6 Mangrove rehabilitation	221
	11.1.7 Mangroves of the Indus Delta: a case study	222
11.2	Seagrasses: benefits and threats	228
11.3	Global climate change	229
	11.3.1 Rise in atmospheric carbon dioxide	229
	11.3.2 Global warming	230
	11.3.3 Sea-level rise	232
11.4	What are mangroves and seagrasses worth?	233
11.5	Have mangroves and seagrasses a future?	236

Further reading	238
References	240
Index	261

1 Mangroves and seagrasses

To a land animal or plant, the sea is a hostile environment. High salinity, wave action, and fluctuating water levels present problems that are rarely experienced in terrestrial or freshwater habitats. Nevertheless, two great assemblages of angiosperms—vascular flowering plants—have overcome these hazards and successfully colonized the sea: mangroves and seagrasses.

Mangroves are dicotyledonous woody shrubs or trees that are virtually confined to the tropics. They often form dense forests that dominate intertidal muddy shores, frequently consisting of virtually monospecific patches or bands. Mangroves stabilize the soil and create a habitat which is exploited by a host of other organisms: through this, and in their role as photosynthetic primary producers, they are the basis of a complex and productive ecosystem. The mangrove trees themselves, and the other inhabitants of the mangrove ecosystem, are adapted to their unpromising habitat, and can cope with periodic immersion and exposure by the tide, fluctuating salinity, low oxygen concentrations in the water, and—being tropical—frequently high temperatures. The total mangrove area in the world is of the order of 18 000 000 ha (Spalding *et al.* 1997).

Seagrasses are monocotyledonous plants, typically with long, strap-like leaves, although in fact they are not true grasses. They may be intertidal or subtidal, down to depths of about 50 m (Duarte 2001). Intertidal seagrasses may be quite small, but subtidal seagrass meadows can comprise quite large plants, physically supported by the water. Like mangroves, they often dominate their habitat and stabilize the sediment in which they grow. A seagrass meadow, like a mangrove forest, creates a physical environment and provides a source of primary production on which a community of other organisms depends. Unlike mangroves, seagrasses are not restricted to the tropics but occur in all oceans and most latitudes other than polar ones. Estimating total area is fraught with difficulty but, worldwide, seagrass meadows probably cover between 16 000 000 and 50 000 000 ha (Green and Short 2003).

Mangroves and seagrasses are the two great assemblages of marine vascular plants. The key to their success lies in their adaptations to their exacting environment. How they survive, the nature of the ecosystems that depend on them, and their wider significance are the subject of this book.

1.1 Mangroves

Mangroves are defined as woody trees and shrubs which flourish in mangrove habitats (or mangals), which is almost, but not quite, a tautology. True, or exclusive, mangroves are those which occur only in such habitats, or only rarely elsewhere. There is in addition a loosely defined group of species often described as mangrove associates, or non-exclusive mangrove species. These comprise a large number of species typically occurring on the landward margin of the mangal, and often in non-mangal habitats such as rainforest, salt marsh, or lowland freshwater swamps. Many epiphytes also grow on mangrove trees: these include an assortment of creepers,

Table 1.1 The principal mangrove species. From Tomlinson (1986) reproduced by permission of Cambridge University Press.

Family	Genus	Number of species
Major components		
Avicenniaceae	*Avicennia*	8
Combretaceae	*Laguncularia*	1
	Lumnitzera	2
Palmae	*Nypa*	1
Rhizophoraceae	*Bruguiera*	6
	Ceriops	2
	Kandelia	2
	Rhizophora	8
Sonneratiaceae	*Sonneratia*	5
Minor components		
Bombacaceae	*Camptostemon*	2
Euphorbiaceae	*Excoecaria*	2
Lythraceae	*Pemphis*	1
Meliaceae	*Xylocarpus*	2
Myrsinaceae	*Aegiceras*	2
Myrtaceae	*Osbornia*	1
Pellicieraceae	*Pelliciera*	1
Plumbaginaceae	*Aegialitis*	2
Pteridaceae	*Acrostichum*	3
Rubiaceae	*Scyphiphora*	1
Sterculiaceae	*Heritiera*	3

orchids, ferns, and other plants, many of which cannot tolerate salt and therefore grow only high in the mangrove canopy.

True mangroves comprise some 55 species in 20 genera, belonging to 16 families. Taxonomy is not now regarded as the most glamorous aspect of biology; nevertheless, these figures indicate an important feature of mangroves as a group. They are taxonomically diverse. From this we can infer that the mangrove habit—the complex of physiological adaptations enabling survival and success—did not evolve just once and allow rapid diversification by a common ancestor. The mangrove habit probably evolved independently at least 16 times, in 16 separate families: the common features have evolved through convergence, not common descent.

The principal mangrove families and genera are listed in Table 1.1. Most families are represented by a small number of mangrove species, and also contain non-mangrove species. However, of the 30 or so species that represent the major components of mangal communities, 25 belong to just two families, the Avicenniaceae and Rhizophoraceae. These families dominate mangrove communities throughout the world.

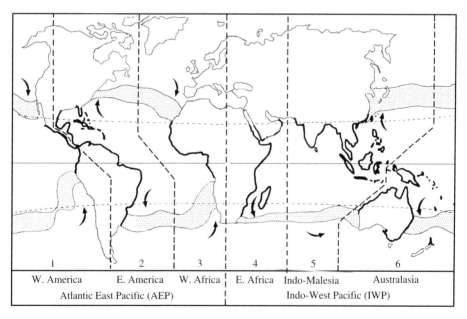

Fig. 1.1 World distribution of mangroves in relation to the January and July 20°C sea-temperature isotherms. The vertical dashed lines indicate the biogeographical areas discussed in Chapter 10, and the arrows are major ocean currents. The horizontal dashed lines indicate limits of the tropics (23.5°N and S). The term Indo-Malesia describes a biogeographic region including India, southern China, Malaysia, Indonesia, and other parts of south-east Asia. Reproduced from Duke, N.C. 1992. Mangrove floristics and biogeography. In *Tropical Mangrove Ecosystems* (Robertson, A.I. and Alongi, D.M., eds), pp. 63–100. Copyright by the American Geophysical Union.

Mangroves are almost exclusively tropical (Figure 1.1). This suggests a limitation by temperature. Although they can survive air temperatures as low as 5°C, mangroves are intolerant of frost. Seedlings are particularly vulnerable. Mangrove distribution, however, correlates most closely with sea temperature. Mangroves rarely occur outside the range delimited by the winter position of the 20°C isotherm, and the number of species tends to decrease as this limit is approached. In the southern hemisphere, ranges extend further south on the eastern margins of land masses than on the western, reflecting the pattern of warm and cold ocean currents. In South America, for example, the southern limit on the Atlantic coast is 33°S. On the Pacific coast, the cold Humboldt current restricts mangroves to 3°40'S.

In Australia and New Zealand, mangroves extend further south, to around 38°S: the southernmost latitude at which mangroves are found is at Corner Inlet, Victoria, Australia, where a variety of *Avicennia marina* occurs at 38°45'S. This extreme distribution may be due to local anomalies of current and temperature, or to the local evolution, for some reason, of an unusually cold-tolerant variety. Possible reasons for the limits in mangrove distribution are discussed in Chapter 2.

The geographical regions indicated in Figure 1.1 correspond to distinct regional differences in the mangrove flora discussed further in Chapter 10.

1.2 Seagrasses

Seagrasses can be seen as more fully adapted to a life in the sea than mangroves, with most being permanently submerged, although some species of *Zostera*, *Phyllospadix*, and *Halophila* grow intertidally (Hemminga and Duarte 2000). The maximum depth at which seagrasses occur is 90 m (Duarte 1991).

Worldwide, seagrasses—like mangroves—do not comprise a large number of species: approximately 50 species in 12 genera. It is unlikely that in the past the total number of species has ever been much greater than this. Again like mangroves, seagrasses are polyphyletic: the seagrass habit appears to have evolved more than once. It is perhaps surprising that there are so few seagrass species: more than 500 species of angiosperm live in fresh water, and hundreds have adapted to saline conditions on land (Duarte 2001).

Conventionally, seagrasses are classified on the basis of their morphology into two families, the Potamogetonaceae and Hydrocharitaceae, each of these being divided into a number of subfamilies. Molecular comparisons

Table 1.2 The seagrasses. Reprinted from Duarte, C.M. (2001). Seagrasses. In *Encyclopaedia of Biodiversity* (ed. S.A. Levin), volume 5 pp. 255–268. Academic Press, with permission from Elsevier.

Family (from morphology)	Subfamily	Genus	Family (from molecular evidence)
Potamogetonaceae	Zosteroidea	*Heterozostera* *Phyllospadix* *Zostera*	Zosteraceae
	Posidonoideae	*Posidonia*	Posidonieaceae
	Cymodoceoideae	*Amphibolis* *Cymodocea* *Halodule* *Syringodium* *Thalassodendron*	Cymodoceaceae
Hydrocharitaceae	Halophiloideae Hydrocharitoideae Thalassioideae	*Halophila* *Enhalus* *Thalassia*	Hydrocharitaceae

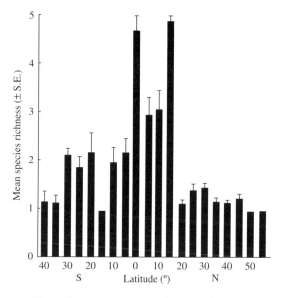

Fig. 1.2 The mean species richness in 596 seagrass meadows in relation to latitude. Reprinted from Duarte, C.M. (2001). Seagrasses. In *Encyclopaedia of Biodiversity* (ed. S.A. Levin), volume 5 pp. 255–268. Academic Press, with permission from Elsevier.

suggest a slightly different phylogeny, with the morphologically based subfamilies promoted to full families (Table 1.2).

Seagrasses are not restricted to tropical or subtropical latitudes, and extend into high northern and southern latitudes, although there is a tendency for more species to be present in the tropics (Figure 1.2).

6 THE BIOLOGY OF MANGROVES AND SEAGRASSES

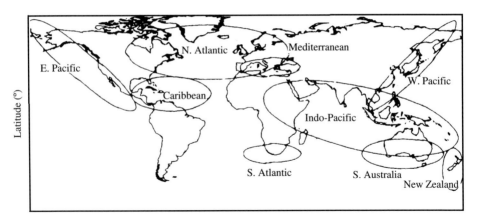

Fig. 1.3 The distribution of distinct seagrass floras. Reprinted from Duarte, C.M. (2001). Seagrasses. In *Encyclopaedia of Biodiversity* (ed. S.A. Levin), volume 5 pp. 255–268. Academic Press, with permission from Elsevier.

Globally, seagrass distributions suggest nine regional floras, as indicated in Figure 1.3:

- temperate North Atlantic
- temperate East Pacific
- temperate West Pacific
- temperate South Atlantic
- Mediterranean
- Caribbean
- Indo-Pacific
- South Australia
- New Zealand

Table 1.3 indicates the distribution of seagrass genera between these regions. Several genera occur in more than one region, as do a few species. The cosmopolitan *Zostera marina*, for example, is found in the North Atlantic, Mediterranean, and both East and West Pacific (Hemminga and Duarte 2000). Possible explanations for the distribution of seagrass genera and species are discussed in Chapter 10.

Table 1.3 The geographical distribution of seagrass genera. Reprinted from Duarte, C.M. (2001). Seagrasses. In *Encyclopaedia of Biodiversity* (ed. S.A. Levin), volume 5 pp. 255–268. Academic Press, with permission from Elsevier, and Hemminga, M.A. and Duarte, C.M. (2000), with permission of Cambridge University Press.

	New Zealand	South Australia	Indo-Pacific	West Pacific	East Pacific	Caribbean	South Atlantic	Mediterranean	North Atlantic	Species/genus
Heterozostera		1								1
Phyllospadix					3				2	5
Zostera	1	3		2	1		1	2		10
Posidonia		8						1		9
Amphibolis		2								2
Cymodocea			3					1		4
Halodule			2			1				3
Syringodium			1			1				2
Thalassodendron		1	1							2
Halophila			8	4		3				10
Enhalus			1							1
Thalassia			1			1				2
Total	1	15	17	6	4	6	1	4	2	51

2 Mangrove trees and their environment

Typical mangrove habitats are periodically inundated by the tides (Figure 2.1). Mangrove trees therefore grow in soil that is more or less permanently waterlogged, and in water whose salinity fluctuates and, with evaporation, may be even higher than that of the open sea. How do they cope?

2.1 Adaptations to waterlogged soil

The underground tissues of any plant require oxygen for respiration. In soils which are not waterlogged, gas diffusion between soil particles can supply this need. In a waterlogged soil, the spaces between soil particles are filled with water. Even when water is saturated with oxygen, its oxygen concentration is far below that of air, and the diffusion rate of oxygen through water is roughly 10 000 times less than through air (Ball 1988a).

Oxygen movement into waterlogged soils is therefore severely limited. Moreover, oxygen that is present is soon depleted by the aerobic respiration of soil bacteria. Thereafter, anaerobic activity takes over. The result is that mangrove soils are often virtually anoxic.

A convenient method of measuring whether a soil is in an anaerobic state is the redox potential (redox being a telescoping of reduction and oxidation). This can be tested by use of a platinum electrode probe, which senses the redox state of the surrounding soil. The redox scale is in millivolts. A well-oxygenated soil will have a redox potential above $+300$ mV. As oxygen availability decreases, so does the redox potential, so that an anoxic mangrove soil may give a reading of -200 mV or lower.

Some unusual bacterially mediated soil chemistry takes place at low soil oxygen levels. As the soil becomes progressively more anoxic and reducing,

Fig. 2.1 Mangroves (*Avicennia marina*) at the edge of a tidal creek in the Indus delta, Pakistan.

bacteria convert nitrate to gaseous nitrogen; this is typical of redox potentials of +200 to +300 mV. As oxygen declines further (redox potentials of +100 to +200 mV) iron is converted from its ferric (Fe^{3+}) to ferrous (Fe^{2+}) form. Since ferric salts are generally insoluble and ferrous ones soluble, this has the effect of releasing soluble iron and inorganic phosphates. These can be used by plants, although excessive uptake of iron is toxic. Finally, with redox potentials of −100 to −200 mV, sulphate is reduced to (toxic) sulphide, and carbon dioxide to methane (Boto 1984). The latter two reactions can result in mangrove mud being extremely pungent, as anyone who has worked in a mangrove swamp can testify, and certainly do not make the environment any more favourable for plant growth.

Mangrove trees have adapted to survive in such unpromising surroundings. The most striking adaptations are various forms of aerial root. The roots of most trees branch off from the trunk underground. In well-oxygenated soil, there is little difficulty in obtaining the oxygen needed for respiration. This is not so in waterlogged soils, and special aerating devices are required. In growing *Rhizophora*, roots diverge from the tree as much as 2 m above ground, elongate at up to 9 mm/day, and penetrate the soil some distance away from the main stem (Figures 2.2 and 2.3). As much as 24% of the above-ground biomass of a tree may consist of aerial roots: the main trunk, as it reaches the ground, tapers into relative insignificance. Aerial

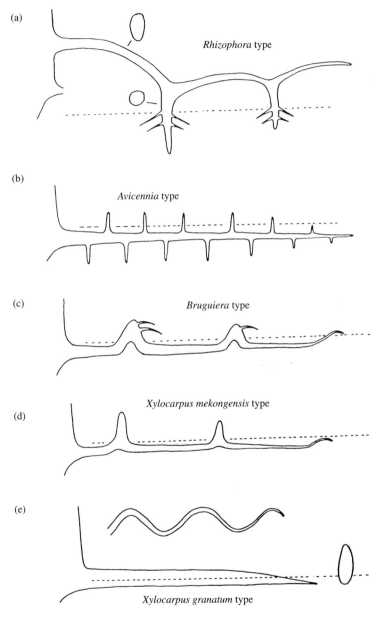

Fig. 2.2 Different forms of mangrove root structure. Reproduced with permission from Tomlinson, P.B. 1986. *The Botany of Mangroves*. Cambridge University Press, Cambridge.

MANGROVE TREES AND THEIR ENVIRONMENT 11

Fig. 2.3 Mangrove forest in western peninsular Malaysia, showing the aerial roots of *Rhizophora*. There is little or no understorey vegetation, except in the sunlit clearing in the background, where it consists almost entirely of *Rhizophora* seedlings.

roots sometimes branch, but apparently only if they are damaged, for instance by wood-boring insects or isopod crustacea (Gill and Tomlinson 1977; Brooks and Bell 2002).

Because of their appearance, and because they provide the main physical support of the trunk, the aerial roots of *Rhizophora* are often termed stilt roots. On reaching the soil surface, absorptive roots grow vertically downwards, and a secondary aerial root may loop off and penetrate the soil still further away from the main trunk. The aerial roots of neighbouring trees often cross, and the result may be an almost impenetrable tangle. This makes life very difficult for the mangrove researcher. Tomlinson (1986) quotes the world record for the 100 m dash through a mangrove swamp: 22 min 30 s.

The method of aerating the underground roots is understood best in the red mangrove *Rhizophora mangle* of Florida. Functionally, the aerial components can be divided structurally into more or less horizontal arches, and vertical columns. These have no problems in achieving adequate gas exchange, at least at low tide. In contrast, the underground roots are in a permanently hypoxic, or even anoxic, environment. The columns have the role of supplying oxygen to the underground roots. Air passes into the

12 THE BIOLOGY OF MANGROVES AND SEAGRASSES

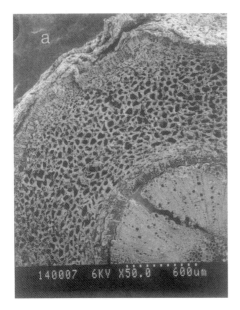 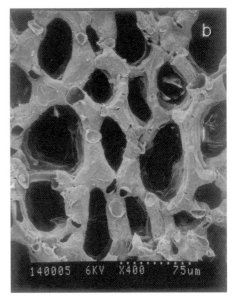

Fig. 2.4 Scanning electron micrographs of aerenchyma tissue near the tip of a pneumatophore of *Avicennia marina*. Original magnifications: (a) ×50; (b) ×400. Micrographs by Fiona Graham, University of Natal. Reproduced from Osborne, D.J. and Berjak, P. 1997. The making of mangroves: the remarkable pioneering role played by seeds of *Avicennia marina*. *Endeavour* **21**: 143–147. ©1997, with permission from Elsevier.

column roots through numerous tiny pores, or lenticels, which are particularly abundant close to the point at which the column root enters the soil surface. It can then pass along roots through air spaces. Roots entering the soil are largely composed of aerenchyma tissue, honeycombed with air spaces which run longitudinally down the root axis (Figure 2.4). These spaces are visible to the naked eye. Their extent and continuity can be demonstrated by blowing down a section of root as through a straw. It is quite easy to blow air through a section of mangrove root up to around 60 cm in length: powerful lungs are required for greater lengths.

The properties of the different components of mangrove roots are illustrated in Table 2.1. Roots in air, with no need to conduct gas internally, contain only a small volume of gas space, equivalent to less than 6% of their volume. In anoxic mud the gas space of a root may be more than half of the total root volume. Between these extremes, roots with different degrees of access to oxygen have intermediate proportions of gas space in their roots. In aerated soil such as drained sand, the need for internal gas movement is slight, and gas space is relatively small. Waterlogged roots have a higher proportion of gas space, unless they are exposed to light. In this case, chlorophyll appears in the surface tissues and photosynthesis offsets the

Table 2.1 Properties of *Rhizophora* roots in different environments. From Gill and Tomlinson (1977), with permission of Blackwell.

Environment	Lenticels	Chlorophyll	Gas space (% volume)
Air	+	+	0–6
Mud	−	−	42–51
Water (light)	−	+	22–29
Water (dark)	−	−	35–40
Drained sand	−	−	22–28

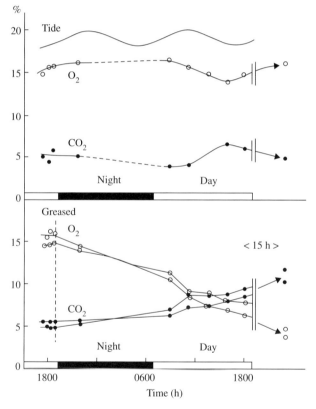

Fig. 2.5 Oxygen and carbon dioxide content of gas sampled from underground root of *Rhizophora*. When the lenticels are occluded with grease (shown by the dashed line on the lower graph) oxygen concentration declines and carbon dioxide rises compared with the controls (upper graph). Time and tidal state (top) are also indicated. Reproduced with permission from Scholander *et al.* 1955. Gas exchange in the roots of mangroves. *American Journal of Botany* **42**: 92–98. Copyright Botanical Society of America.

shortage of environmental oxygen, internal conduction is less important, and internal gas space correspondingly less (Gill and Tomlinson 1977).

The importance of the lenticels for gas exchange has been demonstrated by measuring oxygen and carbon dioxide concentrations in the aerenchyma of *Rhizophora* roots. When the lenticels are occluded by smearing grease over the aerial portion of the root, oxygen declines continuously and carbon dioxide rises (Figure 2.5). Control roots showed fluctuations related to tidal level (Scholander *et al.* 1955).

Looping aerial roots are typical of *Rhizophora*: other species have other forms of root architecture. In *Bruguiera* and *Xylocarpus* a shallow horizontal root periodically breaks the soil surface and submerges again, forming a knee root. In *Xylocarpus granatum* the upper surface of the horizontal root shows above the mud and grows in characteristic sinuous curves (Figure 2.2). A different form of root architecture is shown by species such as *Avicennia*. Shallow horizontal roots radiate outwards, often for a distance of many metres. At intervals of 15–30 cm, vertical structures known as pneumatophores emerge and stand erect, up to 30 cm above the mud surface (Figure 2.2; in *Sonneratia*, pneumatophores may be a staggering 3 m in height). A single *Avicennia* tree 2–3 m in height may have more than 10 000 pneumatophores (Figure 2.6). An abundance of pneumatophores is not, however, an unmixed blessing: because dense pneumatophores slow

Fig. 2.6 Mangroves (*Avicennia marina*) at the edge of the Sinai Desert, Egypt, the most northerly limit of their distribution. The mass of pneumatophores in the foreground belongs to the trees in the background, and is regularly covered by the sea.

water currents and increase sedimentation (see p. 65), they may facilitate their own burial. Mangrove trees appear to balance the advantages and disadvantages of pneumatophore production (Young and Harvey 1996).

Pneumatophores do have some photosynthetic capacity, and may be close to self-sufficient for oxygen during the day, but their primary role is as a conduit for atmospheric oxygen to root tissues. Pneumatophores have abundant lenticels: up to 25 functional lenticels per pneumatophores in *Avicennia* (Hovenden and Allaway 1994). As with the column roots of *Rhizophora*, extensive internal gas spaces allow gas exchange with the underground tissues: aerenchyma may account for up to 70% of root volume (Curran 1985). If the lenticels are sealed with vaseline, oxygen concentration falls and eventually the underground roots asphyxiate.

Simple physical diffusion through the lenticels and along the aerenchyma is probably the main mode of gas movement in mangrove roots, but it may be supplemented by mass flow. One possibility is that changing water pressure during the rise and fall of the tides might alternately compress and expand roots, resulting in successive inhalation and exhalation not unlike that of a vertebrate lung. However, this model predicts that pressure in the roots would increase at high tide and decrease at low. Direct measurements have shown that pressure changes in exactly the opposite way: increasing at low tide and decreasing at high. There is a more convincing interpretation of the observed pressure changes, which might provide a mechanism for the mass flow of air into a root to supplement diffusion. Lenticels are hydrophobic, so that while a root is covered by water they are in effect closed: neither air nor water can enter. Respiration removes oxygen from the air spaces and produces carbon dioxide. Because it is highly soluble in water, the carbon dioxide does not replace the volume of oxygen removed, and gas pressure within the root is therefore reduced. This is confirmed by direct measurement of gas composition in a submerged *Avicennia* root. After a root is covered by the tide oxygen within it falls, carbon dioxide levels do not increase to compensate, and pressure falls. When the tide recedes and the lenticels are again open, air is therefore sucked in (Scholander *et al.* 1955; Skelton and Allaway 1996).

Gas transport by mangrove root systems is so effective that it even aerates the surrounding soil. This has been shown by measuring the redox potential 15 cm below the surface of the soil both close to (<3 cm) and at some distance from (0.5 m) prop roots of *Rhizophora* and pneumatophores of *Avicennia* (Table 2.2). In this case *Rhizophora* clearly made the soil less hypoxic, although *Avicennia* had no significant effect (McKee 1993); in another, similar, study, the opposite was found (Thibodeau and Nickerson 1986).

The relationships between structure and function of mangrove roots are relatively easy to study. Metabolic adaptation of roots to anoxia and waterlogging is less well understood. In aerobic conditions, most organisms

Table 2.2 Redox potentials (Eh; mV) and sulphide concentrations (±S.E.) measured at depths of 15 cm close to (<3 cm) or distant from (approx. 0.5 m) *Rhizophora* prop roots or *Avicennia* pneumatophores. *$P<0.05$, **$P<0.01$. From McKee (1993), with permission of Blackwell.

	Redox potential (Eh; mV)		Sulphide (mM)	
	Near	Away	Near	Away
Rhizophora mangla (prop root)	−80 ± 21*	−203 ± 13	0.17 ± 0.08**	1.70 ± 0.34
Avicennia germinans (pneumatophore)	−154 ± 17	−200 ± 10	0.48 ± 0.15**	1.83 ± 0.26

carry out glycolysis, with the production of pyruvate. This is then disposed of via the Krebs (tricarboxylic acid) cycle and oxidative phosphorylation, with the production of ATP. In anaerobic conditions, this is not possible, and pyruvate is shunted sidewise to form lactate or ethanol. The former is favoured by animals: when aerobic conditions return, the oxygen debt can be repaid and normal oxidative pathways resumed. Ethanol is produced under anaerobic conditions by yeast, a phenomenon that has not gone unnoticed in human society at large.

In experimental conditions, hypoxic roots of *Avicennia* produce a three-fold increase in levels of the enzyme alcohol dehydrogenase, suggesting an adaptive switch to anaerobic metabolic pathways (McKee and Mendelssohn 1987). In *Avicennia* seedlings subjected to 10 days of anoxic conditions significant levels of ethanol accumulated in the tissues. This was true also of *Aegiceras*, *Excoecaria*, and *Rhizophora*, but not of *Hibiscus* and *Bruguiera* (Youssef 1995, cited in Saenger 2002). At least some species of mangrove survive anoxia by metabolic adaptation.

Root architecture differs according to species, but also to some extent adapts to prevailing conditions. In the northern Red Sea, for instance, small areas of *Avicennia* grow in sandy soil at the edge of the Sinai desert. Trees whose roots are regularly inundated by the tide have abundant pneumatophores. On the landward side a few rather stunted trees have contrived to establish themselves in apparently dry sand, where they seem to survive on underground seepage of fresh water. The surface soil is essentially dry sand through which air can readily penetrate. There is no need for special aerating structures: pneumatophores are entirely absent.

A few species of mangrove (*Aegialitis*, *Excoecaria*) lack specialized respiratory roots. In these species the roots lie close to the sediment surface, in a relatively well-oxygenated zone. These species tend to be found in less anaerobic soils (Saenger 2002).

The rooting system of mangroves also has the function of anchorage. The nature of mangrove soil means that roots tend to remain close to the surface, and enter the seriously anoxic depths as little as possible. There are therefore no deeply anchored taproots, but the aerial roots of *Rhizophora* form a combination of guy-ropes and flying buttresses that provide a mechanically effective alternative. The aerial roots of *Rhizophora* and the horizontally spreading system of *Avicennia* provide effective anchorage in a soil that is often fluid and unstable, as well as being anoxic. Compared with most tree species, the roots of mangroves comprise a relatively high proportion of the tree, which will enhance anchorage.

2.2 Coping with salt

Mangroves typically grow in an environment with a salinity between that of fresh water and sea water; sea water comprises approximately 35 g/l salt, (including 483 mM Na^+ and 558 mM Cl^-). This means an osmotic potential of -2.5 MPa, and water must be taken in against this pressure. In some circumstances, mangroves even find themselves in hypersaline conditions, and the problem of water acquisition is correspondingly worse: in the Indus delta of Pakistan, for instance, evaporation raises the prevailing salinity to twice that of the sea. The problem may be compounded by fluctuations in salinity caused by the tide. Variation in salinity may be more difficult to cope with than high salinity itself.

Mangroves deploy a variety of means to cope with this unpromising environment. The principal mechanisms are exclusion of salt by the roots, tolerance of high tissue salt concentrations, and elimination of excess salt by secretion. The interplay between these is complex, and not clearly understood. The requisite field experiments and measurements are not easy to carry out, and laboratory experiments tend to involve isolating one factor and ignoring interactions. Moreover, mangroves in natural situations display an ingenuity that makes interpretation very difficult. Trees that appear to be surrounded by saline or even hypersaline water may be satisfying their requirements from fresh water seepage, or from a subterranean lens of brackish water.

It may seem impossible to tell which of its alternative sources of available water a mangrove tree is actually using. An ingenious series of experiments has exploited the observation that ocean water and fresh water differ in their ratio of two isotopes of oxygen, ^{18}O and the more abundant ^{16}O. The isotope ratio in plant xylem reflects that of the water taken up by the roots. In studies of coastal vegetation in southern Florida, mangroves took water from various sources: in some cases trees whose surface roots were bathed in sea water were relying for their water uptake on brackish or fresh water in the surface layer of the soil. *Rhizophora* seems to depend entirely on water in the top 50 cm of soil, in which 70% of its fine roots are deployed (Sternberg and Swart 1987; Lin and Sternberg 1994).

In *Aegiceras* and *Avicennia*, 90% of salt is excluded at the root surface, rising to 97% as the salinity of the environment increases. The concentration of salt in the xylem sap is about one-tenth of that of sea water (Tomlinson 1986). Although the exclusion mechanism is not understood, it appears to be a simple physical one. Negative hydrostatic pressure is generated within the plant, largely by transpiration processes. This is sufficient to overcome the negative osmotic pressure in the environment of the roots. Water is therefore drawn in. Unwanted ions—and other substances such as dye molecules—are excluded (Moon *et al.* 1986).

The physical nature of this desalination process was demonstrated by Scholander (1955). Decapitated *Avicennia* and *Ceriops* seedlings were positioned in a 'pressure bomb', a container in which the roots, immersed in sea water, could be subjected to pressure by introducing compressed nitrogen while the cut stem was exposed to the atmosphere. With a positive pressure on the roots of 4–4.5 MPa (40–45 atm), the cut stems exuded water at about 4 ml/h, with a salinity of 0.2% NaCl; around 5% of that of the water surrounding the roots at the start of the experiment. As the water exuded over a number of hours was greater in volume than the total water capacity of the seedling, the desalinated water must have come largely from the environment and not from within the seedling itself. In the water surrounding the roots, the salt concentration increased ten-fold, showing that salt was being excluded rather than accumulated in some part of the seedling.

When seedlings were poisoned with either carbon monoxide or the metabolic inhibitor dinitrophenol, the rate of desalination was the same, indicating that the ultrafiltration is a physical process that does not depend on metabolic activity. On the other hand, as external salt concentration increased, the concentrations of sodium and chloride ions in the xylem do not both increase in parallel with each other: with *Avicennia marina*, as the external concentration was increased from 0 to 150% sea water, chloride ions increased five-fold but sodium ions eight-fold. This differential is hard to square with purely physical restrictions on uptake (Tuan *et al.* 1995).

As salt is excluded, it will concentrate in the immediate surroundings of the root, potentially limiting water uptake and constraining mangrove metabolism (Passioura *et al.* 1992).

Even with exclusion of most of the salt, the concentration of sodium and chloride ions within the plant tissues is higher than in non-mangrove species. High salt concentrations are known to inhibit many enzymes, and mangrove enzymes are not particularly salt-tolerant. Intracellular mangrove enzymes are protected by partitioning of solutes within different cellular components. Sodium and chloride ions are at high concentration within cell vacuoles but are largely excluded from the cytoplasm itself. To avoid the consequences of an osmotic imbalance within the cell, high cation concentrations in the vacuole are balanced by high concentrations of non-ionic solutes in the cytoplasm.

Various substances probably contribute, particularly glycinebetaine, proline, and mannitol (Wyn Jones and Storey 1981).

In several mangrove species, among them *Avicennia*, *Rhizophora*, *Sonneratia*, and *Xylocarpus*, sodium chloride is also deposited in the bark of stems and roots. Several species deposit salt in senescent leaves, which are then shed. This may help to remove salt from metabolic tissues. The deciduous mangroves *Xylocarpus* and *Excoecaria* appear to dump excess salt in this way in preparation for a new growing and fruiting season (Hutchings and Saenger 1987).

A number of species of mangrove possess salt glands on their leaves. Leaves of *Aegiceras* and *Avicennia* taste strongly salty when licked, and often carry clearly visible deposits of salt crystals. The lower leaf surface of *Avicennia* is densely covered with hairs, which raise the secreted droplets of salty water away from the leaf surface, preventing the osmotic withdrawal of water from the leaf tissues (Osborne and Berjak 1997).

The glands of *Avicennia* appear to be formed only in response to saline conditions, while in *Aegiceras*, and most other gland-bearing species, they are present regardless of the environmental salinity. Salt glands resemble each other closely across a range of species, presumably as the result of convergent evolution. They also closely resemble glands that secrete other materials, such as nectar, suggesting a possible evolutionary origin. Although the structure of the glands is well known, the secretion mechanism is still not well understood. The secretory cells are packed with mitochondria, suggesting intense metabolic activity. Secretion of salt can be prevented by metabolic inhibitors: it requires energy (Tomlinson 1986; Hutchings and Saenger 1987).

Exclusion, tolerance, and secretion are used with different emphasis by different species, and within a species under different environmental conditions. Table 2.3 summarizes the known occurrence of these mechanisms in a range of mangrove species.

Given the range of methods of coping with salt, it is not surprising that mangrove species differ in the extent of their salt tolerance. Some species, such as *Sonneratia lanceolata*, show maximal growth in solutions between fresh and 5% sea water, whereas most species seem to grow best at 5–50% sea water (Figure 2.7). A few species (*Ceriops decandra*, *Sonneratia alba*) grow very poorly in fresh water, and *Bruguiera parviflora* and *Ceriops tagal* propagules fail to grow at all. The addition of as little as 5% sea water to the culture solution promotes vigorous growth in these species (Hutchings and Saenger 1987; Ball 1988a).

The way in which salinity tolerances and preferences are tested can influence the conclusion. Culture experiments which simply vary salinity levels ignore interactions between salt tolerance and other environmental

Table 2.3 Mechanisms of coping with salt and their known distribution in a range of mangrove species. From various sources.

Species	Exclude	Secrete	Accumulate
Acanthus		+	
Aegialitis	+	+	
Aegiceras	+	+	
Avicennia	+	+	+
Bruguiera	+		
Ceriops	+		
Excoecaria	+		
Laguncularia		+	
Osbornia	+		+
Rhizophora	+		+
Sonneratia	+	+	+
Xylocarpus			+

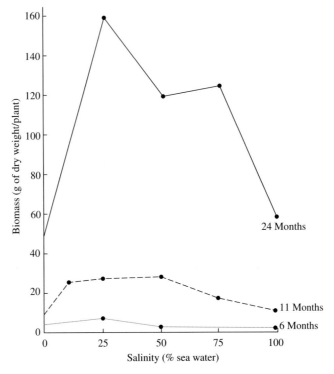

Fig. 2.7 Growth of *Avicennia* seedlings at various concentrations of sea water, measured as dry weight per plant. Reproduced with permission from Hutchings, P. and Saenger, P. (1987). *Ecology of Mangroves*. University of Queensland Press, St Lucia.

variables. For example, salinity interacts with the nature of the substrate: seedlings of *Avicennia* transplanted into sand and exposed to hypersaline conditions (three times the concentration of sea water) for 48 h shed all of their leaves. When the sand was mixed with 5–10% clay, the result was some curling and discoloration of leaves; with a higher proportion of clay in the soil the seedlings appear to survive unscathed. The mechanism of this interaction is not clear, but it is probably connected with ion adsorption on to the clay particles and a consequent reduction in the effective salinity of the soil water (McMillan 1975).

Salt tolerance is also likely to show the same degree of intraspecific variation as any other physiological trait. Individuals collected from different environments, or from different parts of a species' geographical range, may respond differently to varying salt concentrations. Developmental changes also occur: newly released *Avicennia* propagules generally grow best in 50% sea water, but at later stages the optimal salinity is lower (Ball 1988a). If any general conclusion can be drawn, it is probably that mangrove species vary more in the range of their tolerance than in the salinity for optimal growth.

2.3 The cost of survival

Even in conventional plants, the uptake of water and ions is expensive in energy terms. In maize, for instance, about half of root respiration is probably required by the demands of ion uptake, and about 20% of total plant respiration is attributable to uptake and transport costs. In mangroves, coping with a saline environment, these energy costs are likely to be much greater. The leaves of *Avicennia* seedlings respire more rapidly in sea water than in fresh water or 25% sea water. Respiration in the roots falls by 45%, possibly due to the limited supply of photosynthetic products in these conditions (Field 1984; Ball 1988b).

Mangroves also need a relatively greater root mass to satisfy the demand for water. In *Avicennia* and *Aegiceras* the root/shoot ratio is relatively high, and increases with increasing environmental salinity (Figure 2.8; Ball 1988b; Saintilan 1997). Materials committed to root growth cannot be invested in alternative structures, such as leaves.

Once acquired, water must be conserved. Mangrove leaves are often succulent, with thick epidermal walls. The upper surface is often covered with a waxy cuticle and the lower by a dense layer of hairs. These structures—which also, of course, carry construction costs—minimize water loss from the leaves, the main evaporative surface of the tree.

Acquiring and retaining water is therefore energetically expensive in both in construction and running costs. It must be used economically. A plant must assimilate carbon in photosynthesis, which requires the expenditure

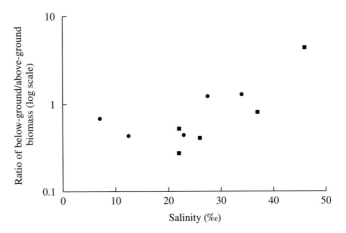

Fig. 2.8 Variation in the ratio of below-ground to above-ground biomass with increasing salinity (parts per thousand, ‰) for communities of *Avicennia* (■) and *Aegiceras* (●). Reproduced with permission from Saintilan, N. 1997. Above- and below-ground biomasses of two species of mangrove on the Hawkesbury River, New South Wales. *Marine and Freshwater Research* **48**: 147–152. © CSIRO Publishing.

of water. A useful index of water economy is the water-use efficiency, the ratio of carbon assimilated to water used. In mangroves this is higher than in comparable non-mangrove trees, particularly in the more salt-tolerant species. Carbon dioxide enters, and water is transpired, through the leaf stomata. At high soil salinities, stomatal conductance—the passage of gases through the stomata—is reduced. This conserves water by reducing transpiration, but also reduces carbon dioxide uptake (and growth). The balance between the two is such that, except under extreme conditions, just sufficient water is expended to maintain the carbon-assimilation rate very near the photosynthetic capacity of the leaf.

Parsimonious use of water leads to other problems. Photosynthesis proceeds most rapidly in *Rhizophora* at a temperature of 25°C, falling off sharply above 35°C. The optimal temperature is typical of the air temperature within a mangrove forest. However, to maximize photosynthesis a leaf must position itself broadside-on to the sun. Maximizing incident light, unfortunately, also maximizes heat gain, and the temperature of a leaf in this position rapidly rises to 10–11°C above air temperature. One way of reducing leaf temperature would be to increase the transpiration rate and lose heat by evaporation. Mangroves cannot afford to do this. Instead, they tend to hold their leaves at an angle to the horizontal, so minimizing heat gain. The angle varies from about 75° in leaves with greatest exposure to the sun, to 0° (horizontal) in leaves in full shade. Cooling is also enhanced by leaf design. Small leaves lose more heat by convection than large ones: leaves exposed to full sunlight, and heat-stressed, are smaller than those that are shaded. Leaves also tend to be smaller in the more salt-tolerant species, where water economy must be more stringent (Ball 1988a; Ball *et al.* 1988). Such constraints on leaf morphology may explain the convergent similarity between the leaves of different mangrove species.

Mangroves must therefore achieve a balance between minimizing water expenditure, holding down leaf temperature, and maximizing carbon dioxide acquisition and growth. The trade-off between growth and salt tolerance is neatly shown by comparing two species of *Sonneratia*, *Sonneratia lanceolata* and *Sonneratia alba*. The former grows in salinities of up to 50% that of sea water, whereas the latter has a broader tolerance and can grow in 100% sea water. Both species show optimal growth at 5% sea water. However, at optimal salinity the growth of *S. alba*, measured as biomass, height, or leaf area, is less than half that of the less salt-tolerant species. A species can apparently opt for salt-tolerance or for rapid growth, not both. An ecological implication of this comparison, of course, is that *S. lanceolata* will be the successful competitor even at a salinity that is optimal for both species, but will not be able to grow or compete at high salinities. This is consistent with the actual distribution of the species along natural salinity gradients (Ball and Pidsley 1995).

The elaborate aerial root structures that enable mangroves to cope with anoxic soils (p. 9) represent construction and maintenance costs to the

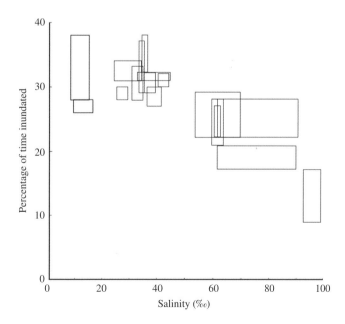

Fig. 2.9 Tolerance range of salinity (parts per thousand, ‰) and percentage of time inundated for 17 Australian mangrove species. From data in Hutchings and Saenger (1987).

plant. In more anoxic soils, more pneumatophores are produced and this investment becomes greater.

Mangroves therefore cope with the environmental stresses of salt and waterlogging, but at the expense of growth, leaf area, and photosynthesis. In extreme conditions, growth may be so restricted that dwarfing occurs. Large areas of the Indus delta in Pakistan, for instance, are covered with dwarf *Avicennia* less than 0.5 m in height; these are not seedlings, but adult trees that are possibly decades old.

Given the costs of tolerance of high salt and low oxygen levels, it is not surprising to find that mangroves trees tend to avoid extremes of both simultaneously. Figure 2.9 shows how a number of Australian mangrove species are distributed in relation to different levels of salinity and waterlogging. Species found at particularly high salinities do not occur at high levels of waterlogging, and vice versa. These limits to distribution are narrower than the extremes that the species could actually survive: the actual distributions reflect interspecific competition as well as physiological tolerance.

2.4 Inorganic nutrients

Plants require an adequate supply of mineral nutrients, particularly nitrate and phosphate. Soluble forms of inorganic nitrogen (except for ammonia) and phosphorus are extremely low in typical mangrove soils, the bulk being present in organic form. Availability to the mangroves depends crucially on the nature of the mangrove soil, and on microbial activity within it.

Where do nutrients such as nitrate and phosphate come from? Possible sources are rainfall, fresh water from rivers or as runoff from the land, tide-borne soluble or particle-bound nutrients, fixation of atmospheric nitrogen by cyanobacteria, and release by microbial decomposition of organic material.

The nutrient contribution made directly by rainfall is probably small, but mangrove areas with heavy rainfall tend to have large freshwater inflow from rivers or land runoff. Potentially, mangroves also have a nutrient source in regular tidal inundations. The relative importance of these sources is not clear, and will undoubtedly vary greatly between different mangrove areas; and, of course, nutrients may be lost as well as gained to river water and the tides. The few cases that have been analysed in any detail indicate that most nutrients available to mangroves are terrestrial in origin: in one study, the contribution from land drainage was between 10 and 20 times that from sea water, depending on season (Lugo *et al.* 1976). Mangrove habitats are more likely to have a net gain of nutrients than a net loss (e.g. Rivera Monroy *et al.*, 1995).

2.4.1 Nitrogen

Fixation of atmospheric nitrogen by root-associated bacteria is an important source of nitrogen for many terrestrial plants. The situation with mangroves is not clear. One method of assessing the nitrogen-fixing ability of soil or root material depends on measuring levels of nitrogenase enzymes, responsible for reducing gaseous nitrogen to ammonia. Nitrogenases also reduce acetylene to ethylene; hence acetylene reduction can be taken as a measure of nitrogenase activity, and of nitrogen-fixing capacity. On this basis, it appears that some nitrogen-fixing ability is associated with the roots of *R. mangle*, *Avicennia germinans*, and *Laguncularia racemosa* from Florida, probably associated with the bacterium *Desulfovibrio* (Zuberer and Silver 1975). Other studies have given more equivocal results, and indeed the applicability of the acetylene-reduction method has been questioned (Boto 1982). Some species of mangrove (*Laguncularia*, *Lumnitzera*, and *Aegiceras* species among them) form leaf nodules containing bacteria. These may well be capable of nitrogen-fixation, as in other angiosperm species, in which case the association represents a further potential source of nutrients to the tree (Hutchings and Saenger 1987).

An intriguing association of a different kind has recently been shown between Caribbean *R. mangle* and the sponges *Tedania* and *Haliclona*, which grow on its roots. Epiphytes and epizoites generally have adverse effects on the mangroves on which they grow, because they block lenticels and impede gas exchange. In this case, it appears that both partners benefit: soluble nitrogenous compounds pass from the sponge to the plant, and carbon compounds in the reverse direction. Sponges grow up to 10 times faster on mangrove roots than on inert substrates, and mangrove roots respond to the presence of sponges with rapid proliferation of rootlets ramifying through the sponge tissue (Ellison *et al.* 1996).

Although there are a few reports of mycorrhizal fungi associated with mangroves, it does not seem that mycorrhizal associations play a major role in mangrove ecosystems. The contribution made by nitrogen-fixing bacteria, and by other organisms such as sponges, is not great; but it may be significant that inter-species relationships seem to revolve round nitrogen availability. This suggests that mangroves are often nitrogen-limited and any additional sources of nitrogen, even though quantitatively small, may be valuable.

The availability of inorganic nitrogen depends also on a complex pattern of activity of bacteria within the soil. Mangrove soil is largely anoxic, apart from a very thin aerobic zone at the surface. Ammonia is produced— by either nitrogen-fixation or decomposition of organic matter—principally in the anoxic zone. It diffuses upwards into the aerobic zone. Although some may be lost to the atmosphere, the bulk is probably oxidized by aerobic bacteria, first into nitrite, then into nitrate ions. This process is termed nitrification. Nitrate then diffuses back down through the anoxic

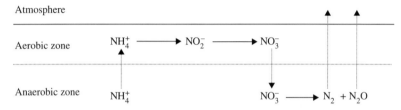

Fig. 2.10 Chemistry of nitrogen compounds in mangrove soil. Solid arrows, conversion; dotted arrows, diffusion. Reproduced with permission from Boto, K.G. 1984. Waterlogged soils. In *The Mangrove Ecosystem: Research Methods* (Snedaker, S.C. and Snedaker, J.G., eds), p. 117, Fig. 7.2. ©UNESCO 1984.

layer. Here its fate depends on circumstances. Nitrate may be taken up by mangrove roots; it may be assimilated by bacteria and immobilized; or it may be reduced by further (anaerobic) bacterial action into either gaseous nitrogen or nitrous oxide. In this last case, the gas is likely to diffuse through the soil and may be lost to the atmosphere (Boto 1982; Hutchings and Saenger, 1987). These processes are summarized in Figure 2.10.

The amount of nitrate actually available to mangrove roots depends on the balance between these processes, hence on the degree of oxygenation of the soil. This, in turn, depends on the inundation régime and on soil composition (p. 8). Soil oxygenation is also affected by animal burrows (p. 126) and by gas leakage from mangrove roots (p. 15), while tannins and other substances exuding from mangrove roots or leaching out of decomposing leaves may inhibit nitrification (Boto 1982).

2.4.2 Phosphorus

The position of phosphorus is just as complex as that of nitrogen. In Australian estuaries considerable phosphate is removed from the water column and adsorbed on sediments as insoluble ferric phosphate. In anaerobic conditions, this is reduced to ferrous phosphate. These processes are impeded by poisoning mangrove sediments with formalin, so must result from metabolic activity of bacteria, and not from physical or chemical processes (Hutchings and Saenger 1987). Ferrous phosphate is soluble, so phosphate may be leached out of the soil and lost. This depends on soil porosity: in fine clay soils interchange between porewater and the water column is less, and such soils may therefore be richer in phosphate and support more flourishing mangroves.

2.4.3 Nutrient recycling

In general, mangrove soils are low in nutrients. In such oligotrophic conditions, mangrove trees are efficient at acquiring and using nutrients.

They minimize nutrient loss by withdrawing a high proportion of nitrogen and phosphorus from senescent leaves: in *Kandelia candel*, this is up to 77 and 58%, respectively (Wang et al. 2003). They may also retrieve nutrients from the soil by proliferation of roots into decaying roots of other trees and old root channels (McKee 2001). Nutrient recycling is therefore very effective. The burying and processing of shed leaves by crabs (p. 104), the nature of mangrove sediments, and the hydrology of mangrove habitats, all encourage nutrient recycling within the ecosystem as a whole (Alongi et al. 1992; Saenger 2002).

2.4.4 Does nutrient availability limit growth?

The generally low levels of nitrate and phosphate in mangrove soils suggest that growth of mangrove trees might be limited by the availability of these nutrients. The best evidence that mangroves are nutrient-limited comes from observing the response to raised nutrient levels. A comparison of two *R. mangle* islands in Florida showed greater mangrove growth on the one with a breeding colony of more than 100 egrets and pelicans, presumably due to the resulting nutrient input in the form of guano (Onuf et al. 1977).

Mangrove productivity has been shown to correlate with soil phosphate levels on a mangrove site in northern Australia (Figure 2.11). In this case, however, the results do not unambiguously prove that nutrient limitation restricts growth. Both tree productivity and phosphate levels also correlate with topographic height of the sample sites. In addition, phosphate exists in different forms, and it is not clear that the method of chemical analysis used to measure soil phosphate (or, indeed, any method of analysis) necessarily reveals the level of phosphate actually available to the trees (Boto 1984).

A more direct approach to testing whether mangroves are nutrient-limited would be to fertilize plots artificially and compare tree growth or productivity with appropriate control plots. This has also been done at a site in northern Australia. Three sites were chosen along a transect across Hinchinbrook Island, Queensland. Two were near the ends of the transect, close to the opposite edges of the island; the third was in the middle of the island, at a more elevated position. One edge site (edge 1) and the middle site were then fertilized with phosphate, the remaining edge site (edge 2) with ammonium. Each site was paired with an untreated control plot, and at all plots, treated and control, mangrove productivity was measured by a pair of litter collectors, for a full year before any treatment, and a full year following the start of treatment. New leaves form as a bud enclosed in a pair of stipules, which are shed as the bud expands. Of the various components of the litter, such as leaves, twigs, and reproductive products, stipule fall correlates most closely with the rate of appearance of new leaves, and was therefore taken as an indication of productivity. The dominant mangroves at all sites were species of *Rhizophora*.

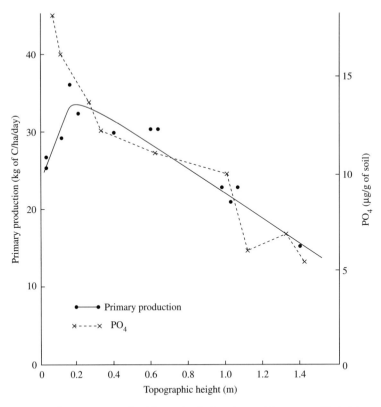

Fig. 2.11 Variation of mangrove productivity and soil phosphate with topographic height on an island in northen Australia. Reproduced with permission from Boto, K.G. 1984. Waterlogged soils. In *The Mangrove Ecosystem: Research Methods* (Snedaker, S.C. and Snedaker, J.G., eds), p. 128, Fig. 7.5. ©UNESCO 1984.

The key results of this study are shown in Table 2.4. The only significant changes were increases in productivity at the middle phosphate-treated site, and the edge site treated with ammonium. The implication is that the mangroves were nitrogen-limited at one low-lying site (edge 2), and phosphorus-limited at the higher (middle) site, but not at the other low-lying site (edge 1). Other studies in Florida and the Caribbean detected phosphate-limited growth of mangroves, but found little evidence of nitrogen limitation (Feller 1995, 1996).

Why the difference between high and low levels at Hinchinbrook Island? More sediment is deposited at lower shore levels, and sediment is probably a major source of phosphate (Boto and Wellington 1983). Lower shores are also inundated more frequently, and for longer, than higher shores. As a result, they are more waterlogged and anoxic; and at more negative redox potentials denitrification will rapidly convert usable nitrate into gaseous

Table 2.4 Leaf production following different nutrient treatments in Hinchinbrook Island, Queensland, Australia. From Boto and Wellington (1983), with permission of Inter-Research Science Center.

Site	Treatment	Stipule fall (g/m²/day)		Difference (%)	Significance
		Before treatment	After treatment		
Edge 1	Phosphate	0.272	0.284	+4.4	n.s.
Control	None	0.409	0.391	−4.4	n.s.
Middle	Phosphate	0.327	0.408	+24.8	$P < 0.05$
Control	None	0.355	0.408	+14.9	n.s.
Edge 2	Ammonium	0.402	0.490	+21.9	$P < 0.05$
Control	None	0.498	0.452	−9.1	n.s.

nitrogen and nitrous oxide, which are lost rapidly to the atmosphere (Figure 2.10).

The mangrove environment, then, is generally poor in nutrients, to the extent that growth of the trees may be restricted. Acquisition of sufficient nitrogen and phosphorus (and, presumably, of other essential substances as well) depends on complex interactions between the soil environment, microbial activity within it, and on its aerobic state.

2.5 Reproductive adaptations

2.5.1 Pollination

The mode of pollination is generally inferred from observation of flower and pollen morphology, and of visits to flowers by possible pollen vectors. Species which produce large amounts of light powdery pollen, which do not smell attractive, and which do not produce nectar, are probably wind-pollinated. In contrast, flowers producing a great deal of nectar to attract animals are likely to be vector-pollinated.

On this basis, *Rhizophora* appears to be wind-pollinated, although bees do visit the flowers regularly, and some nectar may be produced to encourage them. *Sonneratia* is served by bats or hawk moths (see p. 96, Chapter 5 in this volume), large-flowered species of *Bruguiera* by birds, and small-flowered species by butterflies. Bees of various types are probably the main pollinators of *Acanthus*, *Aegiceras*, *Avicennia*, *Excoecaria*, and *Xylocarpus*. *Nypa* may be pollinated either by small bees, or by flies of the family Drosophilididae, related to the familiar fruit fly (Hamilton and Murphy 1988). The flowers of these species are typically clustered, attractively scented, and produce small quantities of nectar, presumably insufficient to attract birds or bats. *Ceriops* and *Kandelia*, and probably other species, are pollinated by a motley collection of small insects (Tomlinson 1986). Pollination can be a chancy business, and in some species it significantly limits fecundity: in *Ceriops australis* and *Rhizophora stylosa*, for example, only 7 and 3% of flowers, respectively, may be fertilized (Coupland et al. 2006).

2.5.2 Propagules

All mangroves disperse their offspring by water. A distinctive feature of the majority of mangrove species is that they produce unusually large propagating structures, or propagules. This term is used because in most mangrove species what leaves the parent tree is a seedling, not a seed or a fruit. After pollination the growing embryo remains on the parent tree, and is dependent on it, for a period that often stretches to many months. The phenomenon is known as vivipary.

Reproduction in orthodox species of plant involves dehydration of the developing seed, accompanied by cessation of DNA synthesis and cell division. These processes seem to be induced by the hormone abscisic acid (ABA). After it is shed, the seed can tolerate desiccation; the process has evolved in a variable environment to allow the embryo to wait, in a quiescent state, for favourable conditions to recur. When conditions are again suitable, the seed germinates, one of the first signs being the outgrowth of roots. ABA also controls the shedding of leaves; it rises transiently in the roots of plants subjected to flooding or salt stress and induces the expression of salt-responsive genes; and it may reduce transpiration water loss by inducing the closure of stomata (Farnsworth 2000).

In most mangrove species which show vivipary, very little ABA is present in the developing propagule, although levels in the rest of the plant remain high. In one species, *R. mangle*, ABA levels in the propagule are normal, but the tissue is insensitive to ABA (Sussex 1975). The propagule does not dehydrate and become quiescent, and DNA synthesis and cell division continue, producing an extended embryonic axis (the hypocotyl) or enlarged cotyledons. Root growth is suppressed as long as the propagule remains attached to its parent. Mangroves differ in this respect from the majority of plants, where seeds are produced which are resistant to desiccation, and which enter a quiescent stage. When the mangrove propagule parts from its parent, there is no period of quiescence, and no ability to resist drying out. Mangrove propagules are poised for germination and root outgrowth whenever the opportunity offers. Conditions may be harsh, but do not show much seasonal variation. In particular, the environment is never short of water, so the strategy of producing quiescent seeds would not be advantageous.

Vivipary and the lack of seed dormancy are therefore linked, through ABA, with resistance of the whole plant to flooding and salt, and these features seem to have evolved in parallel as evolutionarily convergent adaptations to the mangrove environment (Osborne and Berjak 1997; Farnsworth and Farrant 1998; Farnsworth 2000).

The most advanced form of vivipary is shown by *Rhizophora* and other members of the family Rhizophoraceae. After fertilization the embryo develops within a small fruit. As the embryonic axis, or hypocotyl, elongates, it bursts through the surrounding pericarp and develops into a spindle-shaped structure which soon dwarfs the fruit (Figure 2.12). Unlike conventional seeds, there is no period of dormancy, and growth is continuous (Sussex 1975; Tomlinson 1986).

While still attached to the parent, the developing seedling develops chlorophyll and photosynthesizes actively. Water and necessary nutrients are, of course, supplied by the parent. Carbohydrates are also translocated from the parent, and this is probably more important in the later stages of seedling development, as the rate of starch accumulation increases at a time when seedling photosynthesis is actually decreasing (Gunasekar 1993; Pannier and Fraíno de Pannier 1975).

Salt concentrations decline between the pedicel and the cotyledons, and then again between cotyledons and hypocotyl, and decline still further towards the tip of the hypocotyl, reaching levels of less than a third of those found in the pedicel (Lötschert and Liemann 1967). Tissues of the seedling are thus preserved from premature exposure to high salt levels.

The protruding hypocotyl may reach a considerable length. In the New World red mangrove, *R. mangle*, it reaches 20–25 cm and may weigh about 15 g; on *Rhizophora mucronata* of south-east Asia attached seedlings of 1 m have been recorded (Sussex 1975; Tomlinson 1986). Eventually the hypocotyl detaches from the residual fruit and, leaving its cotyledons behind, falls from the parent tree to start its independent life.

Aegiceras, Avicennia, and a number of other mangrove species show a broadly similar form of reproduction, known as cryptovivipary, in which germination and embryonic development take place on the parent tree: in this case, however, the developing hypocotyl fails to penetrate the pericarp. Otherwise, the relationship between parent and offspring is much the same as in *Rhizophora*. The seedling again receives photosynthetic products from its parent, and salt concentrations in the hypocotyl are appreciably lower than those in the adult, or in the intervening tissues of the fruit (Joshi *et al.* 1972; Bhosale and Shinde 1983).

The final group of mangroves reproduce and disperse by more or less conventional seeds, ranging in size from a few millimetres in length to the massive fruit of the aptly named cannonball tree, *Xylocarpus*, which weigh

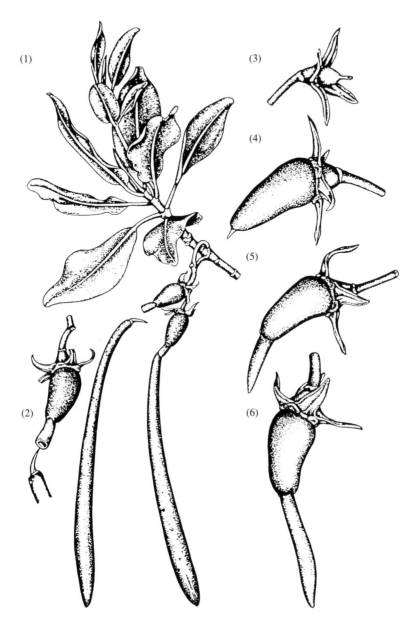

Fig. 2.12 Stages in the development of a *Rhizophora* propagule. (1) Branch with mature propagules, one of which is detached. (2) Details of the relationship between cotyledonary collar and plumule of seedling. (3–6) Stages in elongation of the hypocotyl. The swollen region, proximal to the hypocotyl, corresponds to the fruit, and the hypocotyl to the seedling. From Gill, A.M. and Tomlinson, P.B. 1969. Studies on the growth of red mangrove (*Rhizophora mangle* L.). 1. Habit and general morphology. *Biotropica* **1**: 1–9, with permission of Blackwell.

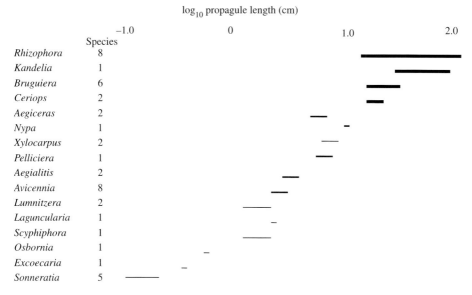

Fig. 2.13 Types of propagule, and their properties, in different mangrove species. Thin bars, non-viviparous; medium bars, cryptoviviparous; thick bars, viviparous. From data in Tomlinson, P.B. (1986).

up to 3 kg and contain up to 20 individual seeds. The different forms of mangrove reproduction, and propagule sizes, are summarized in Figure 2.13. Although most of the studies of ion balance in mangrove propagules and seedlings have focused on the more numerous viviparous and cryptoviviparous species, seedlings of the non-viviparous *Acanthus* also show somewhat lower sodium and chloride concentrations than adult plants. Again, salt appears to be partially excluded from the developing propagule, in this case a seed (Joshi *et al.* 1972).

2.5.3 Fecundity and parental investment

Avicennia in Costa Rica has been estimated to produce more than 2 000 000 propagules/ha per year (Jimenez 1990, quoted in Smith 1992). Translating this into a proportion of total net primary productivity is not easy. Collection of the total litter fall from mangrove trees and its separation into reproductive and other components indicates that the relative effort put into reproductive structures varies widely, according to species and location, but in *Avicennia*, *Rhizophora*, and *Ceriops* it typically ranges between 10 and 40% (data in Bunt 1995). Vivipary and cryptovivipary therefore represent a considerable parental investment.

The parent tree may be able to modulate its investment. In *A. marina*, only about 21% of flower buds survive to form immature fruit, and subsequent mortality means that only 2.9% actually become viable propagules

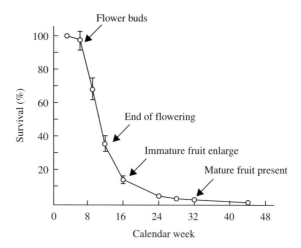

Fig. 2.14 Survival of buds, flowers, and fruits of *Avicennia marina* before dispersal (N = 17 917). Error bars show standard deviations. Reproduced with permission from Clarke, P.J. 1992. Predispersal mortality and fecundity in the grey mangrove (*Avicennia marina*) in south-eastern Australia. *Australian Journal of Ecology* **17**: 161–168, with permission of Blackwell.

(Figure 2.14). Similarly low survival rates have been estimated in other species. Some of the decline can be ascribed to insect attack, but much is due to maternal regulation (Hutchings and Saenger 1987; Clarke 1992; Coupland *et al.* 2006).

2.5.4 Dispersal

The long, pointed appearance of *Rhizophora* propagules hanging on the parent tree has led to the belief that they plummet like darts into the mud below and so immediately establish themselves. This may sometimes happen, but is probably rare. Propagules are often curved, and in any case are not well designed for flight, even vertically downwards. In addition, they may bounce off foliage or aerial roots, or land in water, or on mud that is too hard to penetrate. Besides, this mode of distribution would also result in virtually all seedlings attempting to establish themselves in the shadow of their parent. This is not the most favourable location: some degree of dispersal away from the parent is desirable.

The reality is more complex. *Rhizophora* propagules generally float for some time before rooting themselves. Initially, floating is horizontal. Over a period of a month or so they shift to a vertical position. This makes it more likely that the tip will drag in the mud surface and result in the propagule stranding when the tide recedes. Roots first appear after 10 days or so, and many of the propagules lose buoyancy and sink. Presumably before this has happened the propagule is not ready to establish itself as a seedling. By 40 days virtually all propagules show root growth (Banus and

Fig. 2.15 Rooting and erection of a seedling of *Avicennia*. The intial root growth raises the body of the seedling, minimising contact with the anoxic mud. Reproduced with premission from Chapman, V.J. 1976. *Mangrove Vegetation*. J. Cramer, Vaduz.

Kolehmainen 1975). Most will strand in a horizontal position, and erect themselves after rooting in the mud. Figure 2.15 shows the process of erection in *Avicennia*.

Propagules which do not successfully root after 30 days or so may regain buoyancy and float off again in a horizontal position. They may remain viable for a year or more (Rabinowitz 1978b). Occasionally, propagules are still viable after being transported tens of kilometres inland by hurricanes.

Other mangrove species show broadly similar patterns, with a period of flotation, followed by sinking, rooting, stranding, and establishment as seedlings. An exception appears to be *A. germinans* of Central America, which does not sink and requires stranding before establishing itself. This is an interesting contrast with *A. marina*, the widespread Indo-Pacific species, which sinks on shedding its pericarp (Steinke 1975; Rabinowitz 1978b). Details of the propagules of a number of species are given in Table 2.5.

The timing of these events is affected by circumstances. In sunny conditions, virtually all floating *Rhizophora* propagules pivot to the vertical by 30 days and root within a further 10 days or so; about half of shaded propagules are still floating horizontally after several months. This behaviour will facilitate settling in forest clearings rather than directly under adult trees (Banus and Kolehmainen 1975).

Table 2.5 Characteristics of propagules of a number of Central American mangrove species. F, S, fresh and salt water conditions, respectively. *Avicennia* includes the species *A. germinans* and *A. bicolor*, virtually indistinguishable as propagules; weight and length of *Laguncularia racemosa* represent the means of Atlantic and Pacific populations; dispersal times of *Avicennia*, *Rhizophora*, and *Pelliciera* are times at which 50% had produced roots. Data from Rabinowitz (1978a), with permission of Blackwell.

Species	Weigh (g)	Length (cm)	Floating (days)	Obligate dispersal (days)	Longevity (days)
Laguncularia racemosa	0.4	2.1	23F, 31S	8F	35S
Avicennia spp.	1.1	1.83	Always	~14 F/S	110S
Rhizophora mangle	14.0	22.1			
Rhizophora harrisonii	32.3	36.3	20–100F/S	~40 F/S	>365
Pelliciera rhizophorae	86.2	7.8	1F, 6S	9F, 30S	70S

Salinity also affects dispersal. In several species floating is prolonged in salt water compared with fresh, while *Avicennia* propagules shed their pericarps and sink more rapidly in brackish water (50% sea water) than in either salt or fresh water (Steinke 1975; Rabinowitz 1978b). This presumably makes it more likely that settlement will be in a favourable environment. Immediately after establishment, mangrove seedlings are tolerant of high salinity levels. In *Avicennia*, growth and development are maximal in 50% sea water. Once the seedlings are mature and their reserves are depleted, the optimal salinity is 10–25% sea water (Ball 1988b).

2.5.5 Why vivipary?

Vivipary (including cryptovivipary) is a rare phenomenon among the higher plants: mangroves apart, it occurs largely among plants of tropical shallow marine habitats, such as seagrasses (Elmqvist and Cox 1996). Because many viviparous mangrove species belong to a number of different families, whose other members are not viviparous, vivipary must have arisen through convergent evolution, on a number of separate occasions. This suggests some particular advantage of being viviparous in a tropical tidal, or shallow marine, environment.

It is often implied that the large propagules of many mangrove species enable widespread dispersal of offspring to colonize new areas. Certainly propagules that can float, and remain viable, for periods of weeks or months could in principle travel considerable distances. But do they? The evidence is sparse. Approximately 78% of *A. marina* propagules (which sink after only a few days) were stranded within 2 km of their starting point; most of these are within 500 m. With *Ceriops*, at least 75% remain within 1 m of the parent tree, and 91% within 3 m. In contrast, only 2% of *Kandelia* propagules remain under the parent tree, 12% within 50 m, and the remaining 88% were not recovered and presumed to have dispersed greater distances (Chapman 1976; Clarke 1993; McGuinness 1997). Dispersal ranges probably vary greatly, depending on whether the propagules sink after a few days (*A. marina*) or not (*A. germinans*), or even remain afloat and viable after months in sea water (*Rhizophora*; Table 2.5).

Overall, however, there is little evidence that mangrove propagules regularly disperse over large distances in significant numbers. Nor, in fact, is there necessarily any advantage in scattering offspring thinly across large areas of sea. The chances of a propagule surviving the odyssey, landing in a suitable habitat, and successfully establishing itself there, must be remote compared to its prospects in the habitat from whence it came. The chances are even more remote of two such pioneers settling close enough to have a chance of mutual pollination. The relationship of dispersal with the geographical distribution of mangroves is discussed further in Chapter 10.

It seems more probable that vivipary (and the production of large propagules in general) is an adaptation to the local habitat rather than a stake in the lottery of dispersal. On leaving the parental tree, a mangrove propagule must be physically able to withstand a certain amount of trundling around in the tide and currents. Size may be advantageous. A large propagule which sinks after a few days is likely to have travelled only a short distance, which probably means that it is still within the area in which its parent flourished, which must therefore be reasonably favourable for its own survival, whereas the accelerated settling in sunlight helps to avoid establishment in too shady a position. Finally, the accumulated stores of starch within the hypocotyl mean that rooting, erection of the propagule into a vertical position, and respiration and growth can take place initially without being limited by the rate of photosynthesis. Seedlings of species with large propagules survive better than those with small propagules (Figure 2.16)

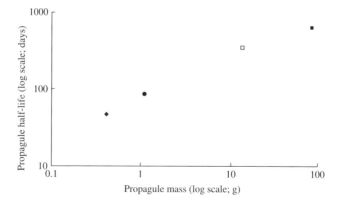

Fig. 2.16 Relationship between propagule weight and seedling survival (half-life) for four mangrove genera. Symbols: ♦, *Laguncularia*; ●, *Avicennia*; □, *Rhizophora*; ■, *Pelliciera*. From data in Rabinowitz 1978b.

and, within a species, larger propagules perform better as seedlings than smaller ones (Rabinowitz 1978c, Sousa *et al.* 2003).

Not all mangrove species show vivipary, and not all have large propagules. Presumably different strategies have been adopted. Perhaps a species with large numbers of small, short-lived propagules exploits gaps in the forest that already exist, while the objective of the long-lived *Rhizophora* propagule is to be around when a new gap appears: dispersal in time, rather than space. Both strategies—and others intermediate on the spectrum between them—are therefore adaptations to an environment that is patchy in time as well as in space.

2.6 Why are mangroves tropical?

As mentioned on p. 4, mangroves are almost exclusively tropical. Geographical ranges are limited by barriers to dispersal, by restricted availability of suitable habitats, or by physiological constraints. Given the pantropical distribution of mangroves, and the numerous muddy estuaries in temperate regions, physiological constraints seem the most likely explanation for latitudinal limits on mangrove distribution.

The key mangrove adaptations are the ability to survive in waterlogged and anoxic soil and the ability to tolerate salt or brackish water. Many, but not all, show vivipary and relatively large propagules. Why should this suite of characteristics be limited by latitude? Plants that tolerate salt, and plants that survive in waterlogged soils, are not unusual in temperate latitudes; in fact, salt-marsh vegetation is as typical of temperate coastal environments as mangroves are of tropical ones. There are temperate trees that cope with waterlogging, or with high salt concentrations. It is only the combination of being a tree, tolerating salt, and coping with waterlogging that is restricted to the tropics.

There are metabolic costs in tolerating salt, and in coping with waterlogging. Are there additional costs in growing and operating as a tree which mean that the mangrove way of life is not feasible where low temperature and short day length reduce photosynthesis for much of the year?

3 Seagrasses and their environment

Seagrasses inhabit a physically challenging environment. They mostly grow in mud or sand, like mangroves, the only significant exception to this being *Phyllospadix*, which can cling to rocky shelves (Hemminga and Duarte 2000). Soft sediments are intrinsically unstable, and seagrasses require an appropriate pattern of growth and root formation that provides adequate anchorage.

Although some species such as *Zostera marina* grow intertidally, the most extensive seagrass meadows are subtidal. Gas diffuses much more slowly in water than in air (p. 8), and while water movement will assist gas exchange for the above-ground components of seagrasses, gas transport within the plant is needed to cater for the underground roots, and suitable adaptations have evolved to achieve this. Water absorbs light, and efficient photosynthesis becomes more difficult to achieve as depth of water increases: in consequence, seagrasses are depth-limited. This is not a problem for mangroves, whose leaves are (mostly) held above the water surface.

Seagrasses must also tolerate high levels of salinity. For subtidal species the problem may be simpler, as salinity, although high, does not fluctuate greatly. Intertidal or estuarine species may need to cope with varying salinity, depending on evaporation and the flow of fresh water from the land.

Finally, for a submerged angiosperm, reproduction presents special problems. The flowers of terrestrial angiosperms are a means of ensuring pollination, either by wind or by animals: neither agent is available in the sea. Both pollination and seed dispersal of seagrasses show special adaptations to their aquatic environment.

3.1 Growth and structure

Seagrasses, like terrestrial monocotyledonous plants, have a rhizomatous growth pattern, shown in Figure 3.1. The horizontal rhizome extends in the sediment, close to the surface. At intervals, it forms a vertical rhizome,

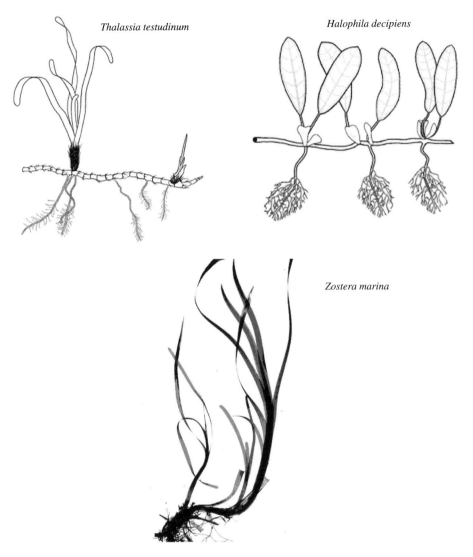

Fig. 3.1 Some species of seagrass, showing the pattern of horizontal rhizome growth. Not to scale: *Halophila decipiens* grows up to 10 cm, *Zostera marina* to around 20 cm, and *Thalassia testudinum* leaves to 30 cm or more. *Thalassia* and *Halophila* drawn by David Crewz, reproduced by permission of Florida Fish and Wildlife Conservation Commission; *Zostera* by permission of Prof. Mike Guiry.

which develops leaves from a basal meristem, and adventitious roots, from which may develop further rootlets. This vertical unit—above-ground leaves and below-ground roots—constitutes a ramet, the modular unit of seagrass growth. The horizontal rhizome, apart from physically maintaining continuity between ramets, also acts as a channel of communication.

Hormone traffic regulates growth, and nutrients are translocated to the growing point of the horizontal rhizome and between ramets: depending on species, up to 40% of the nitrogen required for rhizome growth is translocated from established ramets (Marbà *et al.* 2002).

Seagrasses vary in size, from *Halophila* species, with leaves 1 cm in length, to *Zostera asiatica* and *Enhalus acoroides*, whose strap-like leaves exceed 1 m, and *Posidonia*, whose root may exceed 5 m in length. All seagrasses, however, follow the same general pattern, and 'can be roughly conceived as scaled models of one another' (Hemminga and Duarte 2000), in contrast to the enormous variation in form shown by angiosperms at large.

Extension and branching of the horizontal rhizome results in a seagrass plant expanding, as a clone, to occupy space. Seagrass meadows, as a result, often comprise quite large areas dominated by a single species. In some cases a single clone may continue to expand for many years. Genetic analysis of a *Z. marina* bed in the Baltic showed that it was dominated by a single genotype that included more than 1 million shoots and extended over an area of more than 0.5 ha, making it the largest marine plant recorded. Taking account of the rate of horizontal expansion, the clone may be more than 1000 years old (Reusch *et al.* 1999).

The pattern of horizontally branching rhizomes with anchoring roots at frequent intervals also effectively holds sediment together and reduces its mobility. The effect is enhanced by the reduction of wave or current energy as water passes through the densely packed seagrass leaves. Like mangroves, seagrasses trap and consolidate sediment, having a major physical effect on their substrate.

If the horizontal rhizomes are a means of a plant exploring and extending in space, the vertical components of seagrass ramets are involved in using resources. In nutrient-poor sediments, the root system may grow more extensively, whereas with increasing water depth (hence diminishing light) leaf size tends to increase. The modular seagrass pattern of growth is flexible, and the allocation of biomass to different components—the pattern of investment—reflects the availability of resources in the environment. Typically, 50–60% of seagrass biomass is underground, in roots or horizontal rhizomes.

3.2 Photosynthesis and respiration

To photosynthesize, a plant needs to absorb light and carry out gas exchange with its environment, acquiring carbon dioxide and releasing oxygen. These are more difficult to achieve for an aquatic plant than for a terrestrial one. Light attenuates rapidly with increasing depth, due to absorption by water as well as scattering by suspended particles. Virtually no light of the spectral range necessary for photosynthesis (350–700 nm) penetrates beyond 200 m in pure water. Seagrasses generally require irradiance greater than about 11% of that at the sea surface, which in practice means that they are restricted to a few tens of metres depth (the upper limit is dictated by desiccation; Hemminga and Duarte 2000; Duarte 2002).

In terrestrial plants—and in mangroves—carbon dioxide enters the leaf through stomata, which also allow oxygen release and control transpiration water loss. This arrangement is not suitable for gas exchange under water, and transpiration does not occur in aquatic plants. Seagrass leaves are surrounded by a thin and porous cuticle through which gas exchange (and the uptake of mineral nutrients) takes place.

The organs of seagrasses are penetrated by a network of gas-filled lacunae, comparable with the aerenchyma tissue of mangroves. These lacunae may bring carbon dioxide produced by respiration to the leaves for photosynthesis, and certainly conduct oxygen to the roots for respiration. Seagrass soils are often richly organic, and almost always low in oxygen, and without this means of gas transport, a submerged life would be impossible. Even with gas conductance, seagrasses occupy sediments only where the redox potential (see p. 8) is greater than -100 mV (Terrados *et al.* 1999; Hemminga and Duarte 2000).

Within the leaves, gas spaces also confer buoyancy, helping to raise the photosynthetic apparatus and maximize light capture. Regular transverse septa interrupt the lacunae, providing a damage-limitation system that prevents the catastrophic flooding that might otherwise be a consequence of tissue damage by herbivores (Kuo and McComb 1989).

3.3 Salinity

Seagrasses flourish in a saline environment. Subtidal species live permanently in seawater at around 35‰; seagrasses in estuaries survive in brackish water, whereas in a shallow lagoon or when exposed at low tide, conditions may become hypersaline. Seagrass species are generally euryhaline but in seriously hyposaline (<10‰) or hypersaline (>45‰) conditions, they suffer stress and are likely to become necrotic and die (Hemminga and Duarte 2000).

The situation is rather different from that experienced by mangroves, in which only the roots are exposed to salt water: in seagrasses the whole plant is immersed. Nevertheless, of the three principal mangrove mechanisms for coping with salt (p. 18)—exclusion, tissue tolerance, and salt secretion—only the third has no parallel in seagrasses.

In seawater, the ratio of sodium to potassium ions is around 30. In *Z. marina* leaves it is 1.04, indicating the active exclusion from the leaf of sodium ions and import, in their place, of potassium ions. The Na^+/K^+ ratio within epidermal cells is even lower (0.87) than in other parts of the leaf, which points to further selective exclusion by the cells themselves. The mechanism is not well understood. An ATPase enzyme in the plasma membrane of epidermal cells appears to be involved in ion transportation. In contrast to

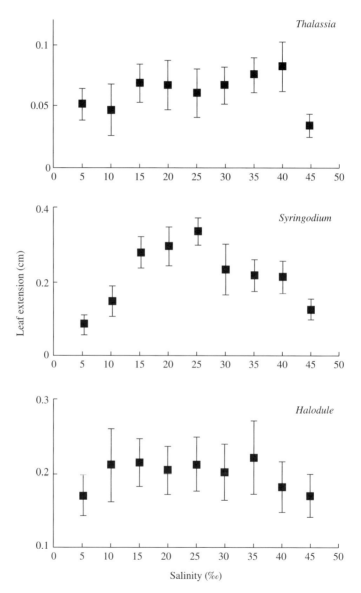

Fig. 3.2 Leaf extension of *Thalassia testudinum*, *Syringodium filiforme*, and *Halodule wrightii* exposed to different salinity treatments for 14 days. Reproduced from Lirman and Cropper (2003), with permission of the Estuarine Research Federation.

the equivalent enzyme in freshwater or terrestrial monocotyledons, this enzyme in *Zostera* is not inhibited by NaCl (Muramatsu *et al.* 2002).

Most intracellular enzymes of seagrasses are not salt-tolerant in this way, and are significantly inhibited by even rather low NaCl concentrations. As in mangroves and other halophytes, sodium ions appear to be sequestered within the cell vacuole and excluded from the cytoplasm. Osmotic balance between vacuole and cytoplasm is probably maintained by organic solutes such as proline and glycinebetaine in the cytoplasm. The concentrations of these fluctuate in parallel with changing salinity, making it possible to maintain more or less constant cytoplasmic levels of sodium (Tyerman 1989).

Although generally euryhaline, seagrass species do vary in their responses to different salinity levels. Figure 3.2, for example, shows the leaf-extension rates of three species of seagrass exposed to different salinity levels. *Thalassia testudinum* has a broad salinity tolerance, with growth reduced only at salinities >40‰, *Halodule wrightii* shows little decline, even at the extremes of salinity tested, whereas *Syringodium filiforme* has the narrowest range of salinity tolerance (Lirman and Cropper 2003).

Different susceptibilities to salinity differences go some way to explain differences in distribution of the different species of seagrass. Salinity tolerance, however, is complex, and species differ not just in how well they survive in a range of constant salinities, but in how they respond to changing salinity. In Biscayne Bay, Florida, *S. filiforme* occurs only in deeper water, not affected by freshwater flow from the land, as would be expected from its more limited salinity tolerance. The more euryhaline species *T. testudinum* and *H. wrightii* are more widely distributed, with *Thalassia* the dominant space-monopolizing species where salinity and other environmental factors are relatively constant, while *Halodule* competes successfully when other species have been removed by disturbance, and remains dominant only in a fluctuating environment (Lirman and Cropper 2003). The relationship between plant distribution and factors such as salinity is often not clear-cut.

3.4 Nutrients

Seagrasses are able to flourish in nutrient-poor environments, requiring only a fraction of the levels of nitrogen and phosphorus needed by other aquatic primary producers such as macroalgae (Duarte 1995). However, growth is often limited by nutrient availability, as shown by the response to artificially supplementing these essential nutrients (Figure 3.3). Seagrasses may be limited by availability of nitrogen, phosphorus, or both.

Availability of nutrients to roots depends on the nature of the soil. Seagrass and mangrove sediments are similar: both tend to be richly organic and anoxic, and the chemical processes with respect to nitrogen and phosphorus

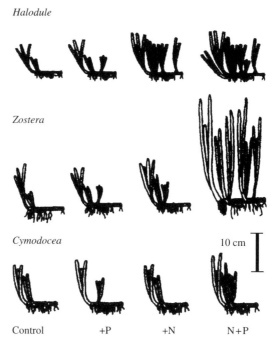

Fig. 3.3 Growth response of *Halodule uninervis*, *Zostera capricorni*, and *Cymodocea serrulata* to fertilization with phosphorus (+P), nitrogen (+N), and both (N + P). Reproduced from Udy, J.W. and Dennison, W.C. (1997). Growth and physiological responses of three seagrass species to elevated sediment nutrients in Moreton Bay, Australia. *Journal of Experimental Marine Biology and Ecology* **217**, 253–277, with permission of Elsevier.

are also similar (p. 24). As with mangroves, nitrogen-fixing cyanobacteria are present, both on the leaf surface of seagrasses, and within the sediment. Nitrogen availability to seagrass plants depends on the degree of anoxia in the soil, and the consequent balance between the various soil nitrogen processes.

Phosphate availability also depends on the composition of the sediment. In carbonate sediments—as where the mineral portion of the sediment derives ultimately from coral reefs—phosphate interacts with carbonate and is less available as free phosphate: in such cases seagrass are likely to be phosphorus-limited. In siliceous sands, growth is likely to be limited by nitrogen, or by nitrogen and phosphorus together (Hemminga and Duarte 2000).

Unlike mangroves, seagrasses can also take up nutrients such as ammonia and phosphate through their leaves. Although nutrient concentration in the water column is generally extremely low, particularly in the tropics, this source of nutrients can be very important. In *Amphibolis antarctica* and *Phyllospadix torreyi*, which may grow on rocks rather than in particulate sediments, leaves are the dominant source. In other species, the importance of leaf uptake varies with circumstances, but even in nutrient-poor tropical waters it can represent the majority of nutrient uptake by a seagrass plant (Stapel *et al.* 1996; Terrados and Williams 1997).

Given the limited availability of nutrients, seagrasses run a tight internal nutrient economy, with nutrients conserved by recycling and transfer from old to young and growing leaves.

3.5 Reproduction

Rhizomatous growth enables seagrasses to expand and occupy space on the sea bed. They may also propagate vegetatively by dispersal of fragments of rhizome, or of above-ground components that retain the ability to form roots (Hall *et al.* 2006). Seagrasses devote relatively little energy to sexual reproduction. Most seagrass shoots bear flowers or fruit very rarely. A few species, however, flower frequently. In *Enhalus acoroides*, flowering is more or less continuous over the year, and represents the investment of around 20% of above-ground production.

Where flowering is seasonal, its timing may be determined by sea temperature, or by tides. Flowers are generally inconspicuous and may be minute: there is no need to attract insect or other animals to transport pollen. For reasons that are not completely obvious, aquatic plants have never exploited animal vectors. Most seagrasses are dioecious, with separate male and female plants. This is true of nine out of 13 seagrass genera, compared with about one in 13 of angiosperm genera overall, and may be a means of avoiding self-pollination.

In the absence of animal vectors and wind, the principal means of pollination in terrestrial plants, seagrasses rely on water transport. Most show hydrophilous pollination. Most commonly, pollen grains are shed by the anthers of the male flower into the water. In some cases the anther may itself be shed, float to the water surface, and shed a rain of pollen grains—denser than sea water—onto female flowers below. In *Enhalus acoroides* the entire male flowers break off, and pollen grains are discharged onto the surface of the water and settle on female plants at low tide. (McConchie and Knox 1989; Hemminga and Duarte 2000).

In many species of seagrass, particularly in the Hydrocharitaceae, the pollen grains are long and thread-like: mature *Amphibolis* pollen grains are a staggering 3 mm long and 0.02 mm in width, much the largest of any angiosperm (McConchie and Knox 1989). In the Potamogetonaceae pollen grains are approximately spherical or ellipsoidal, but even here they often form long threads, either because they are released in groups longitudinally arranged within a mucilaginous tube (*Halophila*), or because premature germination results in a protruding pollen tube (*Thalassia*). The different filiform structures must be the result of convergent evolution, and are probably an adaptation to the problems of pollen transport in turbulent coastal waters (Ackerman 1995).

There is little information on how far pollen disperses from the parent plant, but the distances are probably not large, and pollination dispersal can be a limiting factor in the success of seagrass reproduction. In *Z. marina*, it has been estimated that 95% of pollen falls within 15 m of its source, and successful pollination is directly related to flower density (Ruckelshaus 1996; Reusch 2003).

Enhalus acoroides reproductive success was lower where the seagrass meadow formed separated fragments rather than a continuous canopy. This effect on pollination success was probably due in part to pollen sparsity. However, pollination success of *Enhalus* was enhanced by increased density of all seagrass species in a mixed stand, not just of *Enhalus* itself. This suggests that a major factor in pollination success was probably the greater hydrodynamic effect of a continuous seagrass meadow in reducing current strength and water movement and speed, so increasing the efficiency of pollen trapping (Reusch 2003; Vermaat *et al.* 2004).

3.6 Propagule dispersal

Seagrass seeds vary in structure and dispersal properties. Many are denser than seawater and are released close to the sediment surface, so probably disperse only a very short distance from the parent plant. This is true, for example, of *Cymodocea* and *Halodule*. Others, such as *Enhalus*, have a buoyant seed, and these may travel further. Direct measures of dispersal distances have, however, rarely been made. In *Z. marina* 90% of seeds spread only 30 m from their source (Ruckelshaus 1996). In contrast, the buoyant seeds of *Enhalus acoroides* and *Thalassia hemprichii* are estimated to be able to travel 42 and 23 km, respectively (Lacap *et al.* 2002).

In many species, dispersal is rapidly followed by germination, without an extended dormant phase. *Thalassia* seeds buried in sediment survive only a week or so and no soil seed bank is established. *Zostera* seeds may persist for 1–2 months, and *Cymodocea* for 7–9 months. In contrast, *Halodule* and *Syringodium* seeds may remain dormant for more than 3 years. The presence or absence of a significant seed bank in the soil affects the ability of seagrasses to re-establish themselves following disturbance by, for example, a hurricane or typhoon (Hemminga and Duarte 2000; Rollon *et al.* 2003).

Amphibolis and *Thalassodendron* are viviparous. Fertilized ovules develop into seedlings while still attached to the parent plant. During the period of attachment, which may last up to a year, resources are transferred into the seedling, presumably enhancing its survival prospects after independence. Vivipary is relatively rare in terrestrial plants, but is the rule rather than the exception in mangroves (p. 30). The seedlings, when released, are buoyant and may disperse considerable distances. This is also true of the non-viviparous *Thalassia*, where the buoyant seedling may travel much greater distances after germination than the negatively buoyant seed before germination. The seedling of *Amphibolis* is equipped with a cluster of grappling hooks which anchor it to structures such as the rhizomes of other seagrass plants, and so facilitate settling. Seeds may also travel great distances in the guts of turtles or dugongs (see Chapter 7), although this has not been demonstrated (Kuo and McComb 1989; Hemminga and Duarte 2000).

In many seagrass species, investment in sexual reproduction and seed production is low. Even in species where seed production is high—a production rate of more than 50 000 seeds/m^2 per year has been recorded in *Halophila*—high seed and seedling mortality means that the contribution to a population is probably generally low. Typically more than 98% of seedlings fail to survive their first year (Hemminga and Duarte 2000).

Vegetative reproduction, by rhizome extension, and the dispersal of rhizome fragments, is probably more important than sexually produced seeds in filling in gaps that appear through disturbance, and in extending the area of a seagrass meadow. Dispersal of viable fragments of leaf or rhizome may also occur over longer distances. Fragments of *H. wrightii* remain viable for up to 4 weeks which, at a dispersal speed estimated at around 9 km/day, gives a potential dispersal range of 250 km. Other species are less able to disperse by this means: *Halophila johnsoni* survives only a few days afloat (Hall *et al.* 2006).

3.7 Seagrasses change their environment

Seagrass growth and survival are affected by the physical characteristics of their environment, such as salinity. The reverse is also true: seagrasses alter their environment, often in quite striking ways. They trap and consolidate sediment and deplete soil nutrients. Some species leak oxygen into the soil, thus modifying the conditions in root zone, or rhizosphere, and affecting the survival and growth of other organisms (Enriquez *et al.* 2001). Even more dramatically, the Mediterranean seagrass *Posidonia oceanica* has a major effect on topography: massive reefs of dead rhizome material accumulate over time. Together with altered patterns of current flow, the result is sculpting of the sea floor into banks, escarpments, and crevices. *Posidonia* grows at the rate of only a few centimetres a year, but over several centuries this has a major effect on the underwater landscape (Kendrick *et al.* 2005).

4 Community structure and dynamics

Mangrove and seagrass ecosystems can be considered at different spatial scales. At the level of the individual tree, mangroves show complex adaptive structural and physiological responses to the immediate environment, particularly to physical factors such as salinity. These are discussed in Chapter 2. At a larger scale, local variations in these physical factors help to determine the overall structure of the mangrove forest. Such local variations, and the structure of the forest, are profoundly affected by the environmental setting in which it grows. Finally, the species present are a subset of those available in the geographical region, so over-riding biogeographical factors must be taken into account. These are considered in Chapter 10. As with mangroves, so with seagrasses: individual plants show adaptations to their environment (Chapter 3), and the structure of a seagrass meadow is a function of its setting, and of its biogeographical relationships.

4.1 Mangroves: form of the forest

Any attempt at categorization is to some extent arbitrary and subjective. In the case of mangroves, classification is at best an attempt to impose order on 'a chaotic mess of special cases' (Thom 1984), and some oversimplification is inevitable. There have been many proposals for classifying the manifold forms of mangal, some of them applying quite well in certain parts of the world, but less well to geographically different situations (Woodroffe 1992; Ewel *et al.* 1998). This chapter discusses the general forms of mangrove forest, and the factors that shape them. The principal local factors are shore morphology, the influence of tides and river flow, and the variations in salinity and sedimentation that derive from these.

Many of the largest mangrove areas, particularly in Asia, are in the deltas of major rivers. Examples include the Indus delta of Pakistan, the Sundarbans where the Ganges and Brahmaputra rivers flow into the Bay of Bengal, the

Merbok and Matang deltas of peninsular Malaysia, the Fly River of Papua New Guinea and, in South America, the mangroves of the Amazon delta. Characteristically, they have a low tidal range and strong freshwater flow carrying a considerable load of sediment, much of which is deposited in the mangroves. Before damming reduced its flow, the Indus, for instance, annually delivered more than 100 000 000 000 m^3 of water and 200 000 000 t of sediment to its delta (Chapter 11). Much of this was retained, rather than being flushed out to sea.

A typical large river delta comprises a number of distributaries splitting the land into many finger-like projections. Typically, the major channels shift to and fro as a result of silting, wave action, and other factors, so that the currently active delta is often flanked by abandoned channels which may be penetrated by more saline water. Mangroves fringe these channels as well as the main distributaries and may indeed spread inland further up an abandoned channel than against the direction of flow of an active one. River-dominated settings therefore offer a range of different topographical and physical conditions, affecting species distributions as well as growth and productivity. It is also likely that the river flow will facilitate the outwelling to the sea of material generated within the mangal, and that a riverine mangrove will have an influence on other nearby habitats (Chapter 9).

Tide-dominated estuaries are characterized by a high tidal range over a shallow intertidal zone which can be colonized by mangroves. The tidal water is often full-strength sea water. Wave action is diffused by passage over the shelving intertidal zone. Typically, the river estuary is funnel-shaped, often with longitudinally orientated intertidal shoals or mudbanks. Mangroves often appear as fringing forests along the edges of the estuary, although sometimes they may fringe open coast. Tidal currents can carry a considerable amount of sediment, which may be trapped within the mangal and contribute to the accumulation of sediment. Mangal sediment and other material can also be removed by tidal flows, so the flux of materials is likely to be bi-directional.

The seaward edges of tide-dominated mangals may be exposed to considerable wave action. With relatively high wave activity and low river discharge the result is often a steeper shore. Wave action may rework sediment into barrier islands or sand spits enclosing broad lagoons. These are more common in New World than in Indo-Pacific mangrove areas. Wave-dominated estuaries tend to be unstable, constantly changing in topography. High river discharge and high wave action can combine to form a complex mixture of sand ridges, coastal lagoons, river mouths, and abandoned distributaries providing a range of habitats for mangrove settlement.

On the landward side of fringing mangroves in estuaries are often large areas of mangrove, often termed *basin mangroves* or interior mangroves. These are sheltered from wave action, and often inundated only infrequently by the tides. Salinity varies greatly, depending on the circumstances. In areas of high rainfall, or of substantial groundwater flow, salinity may be quite low. On the other hand, evaporation, and removal of water by the mangrove trees, may combine to raise the salinity, and in some areas the soil may be distinctly hypersaline, possibly as much as twice the salinity of sea water. Because currents, both riverine and tidal, are low and there is little turbulence, a basin forest is likely to represent a sink for nutrients and sediment, rather than a source for export. Riverine, tide-dominated, and basin mangroves represent extremes of a continuum, rather than separate and distinct categories, and many mangals show elements of all three, merging into each other.

In addition to the three major functional categories described above, there are some other settings in which mangroves are found, which do not straightforwardly conform to this classification. Rising sea level has, particularly in parts of Australia, drowned river valleys and produced sheltered features, known as rias, suitable for mangrove growth. A dramatic expansion in mangroves in the Holocene (8000–6000 years ago) followed a sea level rise (Chapter 10). *Overwash* mangroves occur in the Caribbean: *Rhizophora* is commonly found in the form of small islets, completely overwashed by the tides (hence usually with little leaf litter), and often underlain by an accumulation of peaty soil.

On certain low-energy coasts in the tropics there are mangrove environments dominated, not by tidal or river flow, but by the accumulation of *carbonate*. Wave action may be dampened by a coral reef or sand barrier, behind which lime muds and peat (generated by the mangroves themselves) build up. In some cases the drowning of an old coral reef complex allows the growth of an irregular patchwork of mangroves in depressions in the underlying substrate. An example of this is seen in the mangroves of Ras Mohammed, in the Red Sea. These are at the northern limit of the geographical range of mangroves, and are also in the unusual position for mangroves of growing at the edge of the Sinai desert, a virtually unique combination of environmental circumstances.

Scrub mangroves are usually found in extreme environments, where nutrients are limiting or little fresh water is available. Trees are stunted, and frequently no more than 1 m in height. Examples occur in Florida and in the Indus delta of Pakistan. *Hammock* mangroves are found particularly in the Florida Everglades. Relative isolation from rivers or the sea leads to a domed accumulation of organic peat over a depression. In some small islands of the Pacific and West Indies small pockets of mangroves are entirely cut off from the sea. In the Pacific and West Indies these *inland mangroves* are usually in sink-holes or other depressions which may be either rocky or contain an accumulation of muddy sediments. Near Bir Ali, in southern Yemen, *Avicennia* has managed to establish itself in a completely land-locked volcanic crater. In a few cases, local sea-level changes have resulted in mangroves surviving in extraordinary situations. On Christmas Island, in the eastern Indian Ocean, a flourishing stand of *Bruguiera* has been found more than 20 m above sea level, where it may have survived in splendid isolation for 120 000 years (Woodroffe 1988).

4.1.1 Species zonation

Environmental settings are defined in terms of relatively large-scale land form and physical processes, while the functional classification approach shifts the emphasis to smaller spatial scale and to functional aspects of mangrove vegetation. Within these contexts, the different species of mangrove are not randomly scattered, but often occur in discrete and more or less monospecific zones.

Mangrove species zonation can be considered at different scales. On a tide-dominated shore, a clear vertical sequence of species often appears. An example is shown in Figure 4.1. The reality is often not as simple as this stylized diagram would suggest: for example, *Avicennia* often has a bimodal distribution, being abundant near the seaward margin and also some way upshore. A vertical sequence of species may repeat itself on the banks of tidal creeks and rivers, resulting in a complex two-dimensional pattern like that shown in Figure 4.2, typical of mangrove areas of western Malaysia.

At a slightly larger scale, regular sequences of species may also occur with increasing distance up a river, interacting with vertical zones related to tide level. Even larger regional trends may emerge, such as the gradation of species across the vast delta of the Ganges and Brahmaputra, discussed later

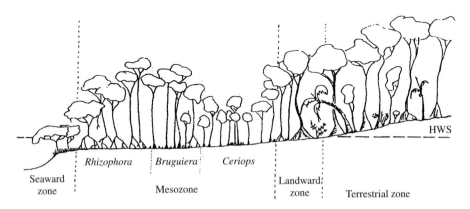

Fig. 4.1 Profile of a shore in north-eastern Australia (an area of high rainfall) to show mangrove zonation. HWS indicates the level of high water, spring tides. Reproduced with permission of Dr N.C. Duke and the publishers from Tomlinson, P.B. 1986. *The Botany of Mangroves*. Cambridge University Press, Cambridge.

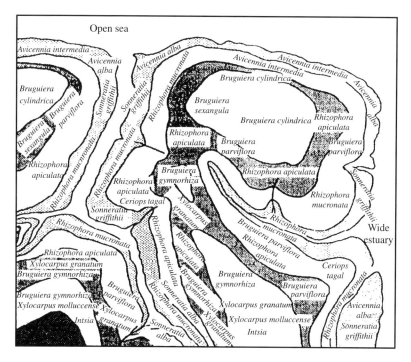

Fig. 4.2 Schematic plan of the complex pattern of species zonation typical of an estuarine mangrove area of western peninsular Malaysia. Reproduced with permission from Smith, T.J. 1992. Forest structure. In *Tropical mangrove ecosystems. Coastal and Estuarine Studies*, no. 41 (Robertson, A.I. and Alongi, D.M., eds), pp. 101–136. Copyright by the American Geophysical Union (adapted from Watson, 1928).

in this chapter (p. 59). Finally, of course, there are major geographical trends in the distribution of species, discussed in Chapter 10.

Not all mangrove areas show clear zonation patterns. Mangroves in Tanzania may be either zoned or unzoned, while those in neighbouring Mozambique are apparently unzoned. In the dwarf mangrove forests of south-eastern Florida the species are randomly intermingled, while in Australia classical zonation patterns appear to be the exception rather than the rule (Snedaker 1982; Smith 1992). The situation is therefore complex and whatever the underlying causes of zonation, they do not appear to apply universally.

Why do mangroves show zonation? Before considering the possible causes of zonation patterns, it is worth remembering that a 'snapshot' view of adult species distribution may be misleading. Whatever the processes that promote zonation, they may operate differently on propagules, seedlings, saplings, and adult trees. The distribution of adult trees is the outcome of all processes that have affected preceding life cycle stages. Additionally, measurement at a specific time, or over a short period, of physical variables thought to be conducive to zonation may not give an adequate picture of the conditions that the adult trees have been subjected to throughout their lifetime. With these provisos, what can be inferred about the causes of mangrove species zonation? Four main approaches to the causes of species zonation have emerged based on concepts from population dynamics, ecophysiology, and geomorphology (Snedaker 1982).

The first suggestion is that zonation results from the physical sorting of floating propagules by water movement, the position of an adult tree within a forest being determined by the point at which a propagule was stranded. Secondly, selection might take place after settlement, for example by different species thriving at different positions along some physical gradient. This suggestion corresponds to the classical paradigm of shore zonation, developed largely on temperate rocky shores. Zonation might be a consequence of gradients created by major geomorphological changes. Finally, zonation might be the product of ecological interactions between species in the community, the sequence of species along a transect in space, for example, corresponding to a natural succession of species with time.

4.1.1.1 *Propagule sorting*

An ingenious explanation for mangrove zonation was put forward by Rabinowitz (1978a), stemming from her observation in Panama that the shore level at which a species occurred correlated with propagule size: the high shore species were those with smaller propagules. She suggested that the distribution of adult trees was based on tidal sorting of propagules. Larger propagules, such as those of *Rhizophora*, were more likely to be stranded and establish themselves at lower and more frequently flooded levels. In contrast smaller propagules, for example those of *Avicennia* or *Aegiceras*, tended to be stranded on the upper shore where they had the necessary time to establish themselves between tidal inundations (Table 4.1; details of propagule size are also given in Figure 2.13). Observations of Indo-Pacific mangrove species also suggested that differences between

Table 4.1 Shore level and mean propagule weight of mangrove species in Panama. *Rhizophora harrisonii* is found on the Pacific coast of Panama, and *Rhizophora mangle* in the Atlantic. Propagules of the two *Avicennia* species (*A. germinans* (both coasts) and *A. bicolor* (Pacific)) were indistinguishable. Data from Rabinowitz (1978b), with permission from Blackwell.

Shore level	Species	Mean propagule weight (g)
Lower	*Rhizophora harrisonii*	32.3±2.6
	Rhizophora mangle	14.0±1.1
	Pelliciera rhizophorae	86.2±6.3
Higher	*Laguncularia racemosa*	0.41±0.02
	Avicennia spp.	1.10±0.11

species in propagule buoyancy and rooting speeds showed some correlation with the shore level at which the species generally occurred as adults (Clarke *et al.* 2001).

However, species found high on the shore are not always those with relatively small propagules, or particular buoyancy properties. Adult *Avicennia* and *Aegiceras*, for instance, are often abundant at low shore levels as well as on the upper shore. Propagule sorting may in some circumstances influence adult distribution, but in general the tides deliver propagules promiscuously across the shore, and adult tree zonation is largely the outcome of post-settlement factors such as physical gradients, selective predation, and competition.

4.1.1.2 *Physical gradients*

The simplest explanation for zonation is that it represents a steady-state response to a gradient in some critical physical variable, and in the case of upshore/downshore gradients it would be that this gradient is maintained by the tidal-inundation régime. In many parts of the world, species distributions on a shore correlate well with tidal-inundation régime. An example is the scheme developed by Watson (1928), based on Port Klang on a mangrove river estuary in western Malaysia, but which has been more widely applied (Table 4.2).

Varying frequency and duration of submergence by the tides creates gradients in several physical variables, including salinity, to which mangroves might respond. Salinity would be an obvious candidate for a gradient-determining species distribution, because of its known effects on individual trees (Chapter 2). Each species, according to this view, would have an optimal salinity range and would be found at the level on the shore where this salinity prevailed. The relationship between shore level and salinity, of course, is not straightforward. In areas with considerable freshwater seepage or runoff, salinity will tend to increase in a seaward direction, while in an arid area with little or no freshwater from the land, and high evaporation rates, the reverse will be the case, and the high shore levels may be significantly hypersaline.

There are often (but not always) reasonable correlations between soil salinity, measured in the field, and mangrove species distributions. Some species, such as *Rhizophora mucronata*, are generally found only in relatively low salinities, below that of normal sea water. Others (for example *Avicennia marina*) can grow where the soil salinity is greater than 65‰, compared with normal sea water whose salt concentration is typically around 35‰ (Hutchings and Saenger 1987; Smith 1992).

Can we therefore conclude that salinity gradients are the principal determinant of species zonation? The hypothesis can be tested experimentally. If zonation results from physiological adaptation to different levels of salinity,

Table 4.2 Watson's tidal-inundation classes. Based on Port Klang, western peninsular Malaysia. Heights have been converted to metric units. From Watson (1928).

Inundation class	Inundated by...	Height above datum (m)	Times flooded per month	R. mucronata	Avicennia	Sonneratia	Ceriops	R. apiculata	B. parviflora	B. gymnorrhiza	Lumnitzera	Acrostichum
1	All high tides	0–2.4	56–62	+								
2	Medium high tides	2.4–3.4	45–49	+	+	+						
3	Normal high tides	3.4–4.0	20–45	+			+	+				
4	Spring high tides	4.0–4.6	2–20						+	+	+	
5	Abnormal (equinoctial) tides	>4.6	<2					+		+		+

then propagules transplanted to a range of different shore levels should succeed best in the same zone, at the same salinity, as their parental species.

The outcome of this experiment is not as predicted. Propagules of four species of mangrove were planted into both low and high intertidal forests differing in inundation frequency and salinity. Survival and growth were then monitored. All four species had better survival rates in the high forest, irrespective of where they had come from, and three out of the four species showed higher growth rates here. The growth rate of the fourth species did not vary between sites. Only one of the species performed better at the level at which it was naturally most abundant; even so, it was out-performed here by the other three species. The results are summarized in Table 4.3. This does not indicate that salinity is irrelevant, but that it is not the key factor in mangrove species zonation. Other similar experiments lead to the same conclusion.

Laboratory experiments suggest that most mangrove species grow best at relatively low salinity. Rather than having different salinity optima, they differ in the range of salinity that they can tolerate. This is shown schematically in Figure 4.3, which also indicates a further important feature of mangrove responses to salinity: the more euryhaline species (those that tolerate a wider salinity range) have in general lower growth rates, *even at their optimum salinity*, than the stenohaline species which are restricted to a narrow range of salinity. The price of wide tolerance is slow growth. The result is zonation along a salinity gradient, because a slower-growing euryhaline species is out-competed even where it would do best, and grows in less favourable salinities where more rapidly growing species are at a relative disadvantage. Zonation results from an interaction between competition and salinity.

Although there are numerous accounts of zonation as a response to physical gradients, one question is rarely raised, probably because no simple answer is apparent. Gradients vary gradually and continuously: why, then, do species form distinct bands with often sharp boundaries? This may also be an effect of interspecific competition and mutual exclusion from the interface where that competition is strongest.

Table 4.3 Growth and survival of propagules of four mangrove species at different shore levels. L, low tidal forest (high inundation frequency, low salinity); H, high tidal forest (low inundation frequency, high salinity). After Smith (1987a).

Species	Normal distribution	Best survival	Best growth
Avicennia marina	L	H	H
Bruguiera gymnorrhiza	L	L/H	H
Ceriops australis	H	H	H
Rhizophora stylosa	L	H	H

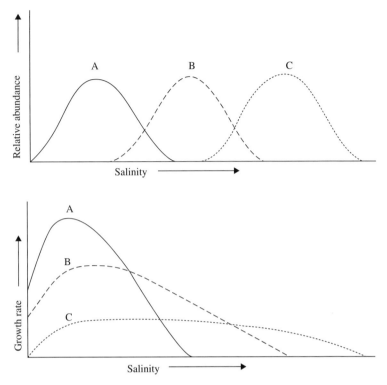

Fig. 4.3 Ecological and physiological responses to a salinity gradient. The upper panel shows the distribution of species A, B, and C along a salinity gradient, with each species dominant over a different salinity range. The lower panel shows the physiological responses of the same species to the salinity gradient. All three species have similar optima, but differ in their growth rate and range of tolerance. The result is that species A dominates at salinities around the optimum by out-competing its rivals. Species B and C are dominant at salinities which are sub-optimal for their growth but which prevent or limit the growth of other species. Reproduced with permission from Ball, M.C. 1988a. Ecophysiology of mangroves. *Trees* **2**: 129–142, with kind permission from Springer Science and Business Media.

Salinity gradients have been taken as an example. Other physical gradients could also contribute. Mangroves differ in their response to variation in other physical factors, such as soil anoxia (Chapter 2, p. 8), soil pH, or sediment composition. Where these factors vary in line with salinity (as is often the case) a consistent pattern of physically determined zonation is more likely to emerge. Where they vary independently of each other the outcome, in terms of patterns of species distribution, may be more complex. This will be particularly true where there are interactions between physical variables. Mangroves appear to make trade-offs in their physiological responses to soil waterlogging and salinity (p. 23 and Figure 2.9), and respond differently to salinity when soil textures differ (p. 21).

4.1.1.3 *Geomorphological change*

Shores consisting of soft sediment are dynamic. Erosion by current or wave action opposes accretion by sediment deposition: if these balance each other, a shoreline may be stable. If they are not in equilibrium, the shore either recedes or extends. Within a delta, distributary channels are apt to shift their course and sometimes become blocked and are replaced by new channels. All of these factors will affect physical gradients and any aspects of mangrove distribution that are affected by them.

In addition, the sea level may rise or fall. This may have either local or global causes. Locally, compaction and consequent subsidence of sediments has the effect of raising the level of the sea relative to that of the mangroves. At a larger scale, tectonic movements may raise or lower land levels. The inland mangroves of Christmas Island, now 20 m above current sea level, are an example (see above, p. 52). Another example is seen in the mangroves of the Sundarbans, at the north end of the Bay of Bengal. Tectonic action has raised the north-western part. Subsidence, due mainly to sediment compaction, has lowered it in the east. The resultant tilting, coupled with the massive input of fresh water in the delta of the Ganges and Brahmaputra to the east, has produced an overall salinity gradient, with the western delta much more saline. The result is an east–west gradient in the distribution of mangrove species, with less salt-tolerant species such as *Heritiera*, *Sonneratia*, and *Nypa* being more common in the east. This east–west gradient in species distribution interacts with the pattern of zonation parallel to the shoreline (Blasco *et al.* 1996).

Global changes in the amount of water in the oceans of the world also alter sea level locally. Such eustatic sea-level changes affect the nature of mangrove zonation. The global impact of sea-level change on mangrove communities is discussed further in Chapter 11.

If change is slow, mangrove communities will have time to adjust, and zonation and other distributions will reflect current conditions. If, however, the change takes place more rapidly, over a period of the same order as the lifespan of individual trees, an observed zonation pattern may have resulted from conditions that no longer obtain. Interpreting patterns in relation to current conditions may be misleading. Figure 4.4 gives an idea of the timescale of relevant geomorphological processes, and of the features of mangroves with which they might interact (Woodroffe 1992).

4.1.1.4 *Plant succession*

An alternative interpretation of the zonation pattern of mangrove forests is that the zones represent stages in a successional sequence, from pioneer species colonizing the seaward fringe to mature climax forest behind. The sequence in space thus reflects an actual or potential sequence in time. Implicit in this model is the idea that each species so modifies the

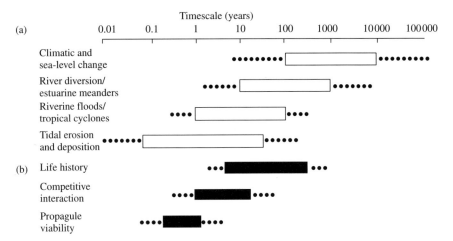

Fig. 4.4 Approximate timescale (shown as a log scale) over which various physical and biological processes operate. Reproduced with permission from Woodroffe, C. 1992. Mangrove sediments and geomorphology. In *Tropical mangrove ecosystems. Coastal and Estuarine Studies*, no. 41 (Robertson, A.I. and Alongi, D.M., eds), pp. 7–41. Copyright by the American Geophysical Union.

environment as to make it possible for a successor species to settle and, in due course, to displace its predecessor. In its extreme form, this theory envisaged mangroves actually creating land: pioneer species trapping mud, consolidating and extending the shoreline, and virtually marching out to sea leaving behind them a trail of successor species culminating in a climax terrestrial forest. Specifically, *Rhizophora mangle* in Florida was credited with the creation of some 600 ha of land over a period of 30–40 years (Snedaker 1982). This successional progress was thought to be countered by disturbances, such as storms (see Chapter 11).

The idea that mangroves create land is now discredited (although it may be unfair to conclude that it is merely 'part of arm-chair musings of air-crammed minds'; Egler, quoted in Smith 1992). Mangroves undoubtedly trap sediment (see p. 64). In so doing, they may accelerate accretion, or retard erosion, of a shoreline. Many mangrove shorelines have mudflats on the seaward side of the mangrove belt: here land has been created by sediment deposition without any part being played by mangroves. Conversely, it is not unusual to find mangrove trees literally falling out of an eroding shore. Accretion can be far faster than mangroves can exploit; and erosion faster than they can counter it. It is much more likely to be the case that mangroves opportunistically follow a shoreline that is accreting for physical reasons, or retreat as a shore erodes than that they make the running.

Another problem with the idea of a seaward succession is that zonation patterns vary a great deal from place to place, and may show inconsistencies such as the bimodal distribution of *Avicennia* (see above). This is hard to reconcile with the implication that zonation patterns simply recapitulate successional sequences.

This does not mean, of course, that mangrove forests do not show patterns of succession, only that succession is not the principal basis of zonation. Perhaps the most convincing account of succession comes from a remarkable long-term study of the mature mangrove forest of Matang, western Malaysia. The Matang is a textbook example of a sustainably managed *Rhizophora apiculata* forest, and is discussed more fully in Chapter 11. Within the Matang area, plots were set aside by the Forestry Department in 1920 in order to determine growth and recruitment rates of trees under different thinning regimes; since 1950 every tree has been tagged and regularly measured, and natural changes in the plots monitored.

Since 1950, the pattern that has emerged is of occasional tree death leaving gaps in the canopy, and of these gaps being rapidly invaded by small plants such as the mangrove fern *Acrostichum*, the woody vine *Derris*, and the shrub *Acanthus*. *Acrostichum* has a particular penchant for growing on the mounds thrown up by the mud lobster *Thalassina* (p. 125). These shrubby colonizers are followed by species of *Bruguiera*. *Bruguiera parviflora* has small and easily dispersed propagules which can form dense seedling patches under a *Rhizophora* canopy: further growth is suppressed unless a canopy gap appears. Subsequently *Bruguiera gymnorrhiza*, a slower-growing and longer-lived shade-tolerant species, becomes more abundant. Overall, species richness has increased with time, typical of woodland successional sequences (Putz and Chan 1986).

There has also been a tendency for shade-tolerant species to increase with time. The first invaders of a patch are more likely to be those that do less well beneath the forest canopy and which need an open space to acquire enough light. Shade intolerance may also explain vertical zonation patterns: at the seaward fringe of a dense mangrove forest there is often a narrow zone of *A. marina*. This species is less tolerant of shade than many of its competitors, and its presence at the edge of the forest may reflect this, rather than a preference for a low shore position (Smith 1987b).

The structure of a mangrove forest is therefore in part explainable in terms of patch dynamics: of gaps appearing by chance and being filled by a changing assemblage of species differing in composition (at least for a time) from the surrounding forest. Eventually something similar to the surrounding forest emerges. With a high incidence of gaps, a mangrove forest could be seen as a mosaic of patches of different successional age: if patches appear relatively rarely, the effect would be of transient aberrations in an otherwise homogeneous, or consistently zoned, environment. The distribution of species that is actually found on a particular shore is therefore the outcome of an interplay between biotic and abiotic factors.

4.1.2 How different are mangroves from other forests?

Succession is part of the normal dynamics of many forest types: the chance appearance of gaps, rapidly colonized by opportunistic 'weeds' that are progressively ousted by slower-growing but more competitive species until a mature canopy forest reappears. Are mangrove habitats, then, just the same as other tropical forests (only wetter)?

Mangrove forests may in general be more subject to disturbance than other forests. As well as the normal disruptions that forests are prone to, mangrove forests may have to contend with typhoons, coastal erosion, wildly fluctuating river discharges, and the tendency of delta channels to wander erratically to and fro. These fluctuations are superimposed on the more or less constant and predictable stresses of high salinity and soil anoxia, which also retard or reverse successional changes. Frequent change, coupled with environmental stresses, makes it less likely that the processes of succession will culminate in a recognizable climax community (Lugo 1980).

Some of the differences and similarities between mangroves and their non-mangrove counterparts are shown in Table 4.4. The comparisons suggest that mangroves resemble (*r*-selected) pioneer species in their reproductive characteristics, but as adult trees they behave more as mature-phase competitive (*K*-selected) species. This observation, that mangroves contrive to have their cake and eat it (Tomlinson 1986), should prove a fruitful insight into the dynamics of mangrove forests.

Mangrove forest, compared with neighbouring tropical rain forest, is poor in species. This would be expected from an ecosystem in which the processes of succession—which include species accumulation—are inhibited. Even allowing for this, the number of plant species is low, and even lower locally, given the tendency for mangroves to grow in relatively monospecific stands within a forest. One reason may be that the species that occur in a particular mangrove forest must be a subset of the species of mangrove found in the region. The total number of species in a regional species pool, even in relatively species-rich regions such as south-east Asia, is not large. (By the standards of tropical rain forest, even the *global* pool of mangrove species is pitifully small.) If species are not present in the geographical area, they cannot be present in a particular mangrove forest. The biogeography of mangroves is discussed further in Chapter 10.

Mangroves differ from other forests in one other marked respect: they virtually lack any understorey vegetation (Figure 2.3). There is often a patchy understorey of mangrove seedlings, but only rarely a significant development of understorey vegetation of other species. Almost the only exception to this is the mangrove fern *Acrostichum*, but even this is not a true understorey plant, as it depends heavily on patches of direct sunlight. Possibly shrubs cannot grow because of the combination of low light levels

Table 4.4 Comparison of the features of mangrove trees and community with pioneer and mature-phase terrestrial forest communities. Modified from Tomlinson (1986), with permission.

Character	Pioneer terrestrial species	Mature-phase terrestrial species	Mangrove species
Trees			
Propagule size	Small	Large	Variable (often large)
Propagule number	High	Low	Often high
Dispersal agent	Often abiotic	Usually biotic	Abiotic (water)
Dispersibility	Wide	Limited	Wide
Seed production	Continuous	Discontinuous	Sometimes continuous
Seed dormancy and viability	Long	Short	Variable
Seedlings	Light-demanding; not dependent on seed reserves	Not light-demanding; dependent on seed reserves	Light-demanding; dependent on reserves
Reproductive age	Early	Late	Early
Life span	Short	Long	Long?
Leaf size	Often large	Medium/small	Medium
Leaf palatability	High	Low	Low
Wood	Soft, light	Hard, heavy	Hard, heavy
Crown shape	Uniform	Varied	Uniform
Competitiveness	For light	For many resources	Mainly for light
Pollinators	Not specific	Highly specific	Not specific
Flowering period	Prolonged or continuous	Short	Prolonged or continuous
Breeding mechanism	Usually inbreeding	Usually outbreeding	Usually inbreeding
Community			
Species richness	Low	High	Low
Stratification	Low	High	Low/absent
Size distribution	Even	Uneven	Even?
Large stems	Absent	Present	Absent (except on old undisturbed stands)
Undergrowth	Dense	Sparse	Usually absent
Climbers	Few	Many	Few
Epiphytes	Few	Many	Few

under the forest canopy, high salinity, and anoxic soil. Whatever the reason, the lack of understorey vegetation, and the relatively low diversity of tree species, mean that mangrove forests are relatively simple in physical structure. Physical complexity also tends to reduce as salinity increases (Janzen 1985).

Mangrove forests are therefore not just tropical forests that happen to live in salt water and anaerobic soil. They have special features of their own: relatively few species and a simple physical structure, and often marked zonation as the consequence either of a frustrated successional sequence, of the effects of geomorphological change, or of physiological responses to clinal physical variables. These, of course, are not mutually exclusive.

4.1.3 Do mangroves make mud?

The distribution and growth patterns of mangroves can largely be interpreted as physiological responses to variations in their physical environment, modulated by competitive interactions between the trees themselves. As the physical environment affects the trees, so they in turn alter their environment. A mangrove forest is not just a mudflat in which trees happen to be growing: the trees affect their substrate. Do they also create it? The question of whether mangroves create land has already been briefly mentioned (p. 60). Although this appears not to be so, established mangrove forests do trap sediment particles, hence accelerate accretion and, conversely, retard erosion. The dynamics of sediment in mangrove forests has been studied in only a few cases, and is far from being clearly understood. Tidal currents, wave action, river flow, salinity gradients, and the topography of mangrove habitats all interact and affect sedimentation in various complex ways.

Many mangrove areas are estuarine, and most large rivers carry a considerable sediment load (see p. 50). As the river water passes through its mangrove estuary, the sediment it carries either settles or remains in suspension and is carried out to sea. A key variable in deciding the outcome is settling velocity, the rate at which particles will sink in still water. If the settling velocity is low in relation to current speed and turbulence, particles will remain suspended. In fresh water, the settling velocity of fine suspended clay particles is low. In the Fly River estuary, a mangrove area in Papua New Guinea, for instance, settling velocity is typically between 10^{-2} and 10^{-1} cm/s. However, when fresh water mixes with salt in the estuary, the suspended particles start to flocculate and form larger aggregates whose settling velocity may be an order of magnitude higher. Sometimes small planktonic animals become trapped by flocs and carried to the mud surface. Flocculation can start at salinities as low as 1‰, about 3% of the concentration of normal sea water. The reasons for flocculation are complex, and relate to the effects of ions on the electrostatic charge on the particle surface, as well as to biological activity: the effect, however, is to promote sinking. The effect may be locally enhanced by salt exuded from mangrove pneumatophores (Wolanski 1995; Young and Harvey 1996).

Suspended particles are also carried in incoming tidal water, and tidal sediment can also be trapped in the mangrove forest. Here, too, the process is complicated and far from fully understood, and several factors are involved. Tidal currents move rapidly up tidal creeks, then spread out over the mangrove forest floor. The flow rate is much greater in the creek than in the forest: in one mangrove system in northern Australia, current velocity was measured in both creeks and forest. In the former it was typically greater than 1 m/s, whereas in the latter it seldom rose above 0.1 m/s. There are two main reasons for this. In the first place, the area of forest may be large in relation to the creek, hence the water moves more slowly when it spreads out. Secondly, and more significantly, the forest offers resistance to water flow. Tree trunks, roots, and pneumatophores add to the friction already created by the mud surface.

Small-scale turbulence around roots keeps particles in suspension when the tide is advancing relatively fast. As the tide advances, particles are carried upshore. At slack water high tide, these particles can sink. The retreating tidal current is too low to resuspend and remove them. The result is net accumulation of sediment from the sea: some 80% of suspended sediment brought in from coastal waters is trapped in mangroves in this way (Furukawa *et al.* 1997).

Net sedimentation rates in mangrove forests vary: measurements of up to 8 mm/year have been made, although most estimates are lower than this (Woodroffe 1992). Sedimentation rate correlates with pneumatophore density in an *Avicennia* forest. It also correlates with the density of simulated pneumatophores, apple tree cuttings of appropriate size planted in bare mud at various densities (Young and Harvey 1996). This confirms that mangrove trees do increase the rate of sedimentation, and thus actively contribute to their environment.

As sediment builds up round *Avicennia* pneumatophores, gas exchange and root respiration will be progressively restricted. This tendency may eventually lead to the death of the tree and either soil erosion or, possibly, its replacement by a successor species. Alternatively, it may be that *Avicennia* can adjust pneumatophore growth to balance the enhanced sedimentation (Blasco *et al.* 1996; Young and Harvey 1996).

Mud is of course a crucial component of the mangrove ecosystem. The mud surface may be the site of significant photosynthesis by unicellular algae and blue-green bacteria. Below the surface bacteria and fungi decompose the organic components of the mud. Many animals also burrow beneath the surface: these may or may not emerge on the surface at low tide. Mud-dwelling animals include a host of deposit or filter-feeders, detritivores, herbivores, and predators. They are discussed in Chapter 6.

As well as contributing to the mud, mangrove trees provide a hard substrate on which other organisms can grow. Pneumatophores, aerial roots, and even the lower branches and leaves, are often festooned with algae, or covered with barnacles and oysters. These in turn are fed on by a host of predators. Parts of the mangrove trees that lie beyond the reach of the tides represent an environment not greatly different from that of a terrestrial forest, occupied by an essentially terrestrial fauna of insects, mammals, and birds. These invaders from the land are described in Chapter 5.

Finally, of course, mangroves are a major source of primary production. Particularly in the form of leaf litter, but also reproductive products, twigs, and eventually entire dead trees, this dominates energy flow pathways, directly or indirectly fuelling virtually the entire ecosystem.

The contribution of mangrove trees to the structural complexity and productivity of their environment is obvious. They also affect the environment more subtly. The problems of maintaining aerobic roots in anoxic mud were discussed in Chapter 2. Sub-surface transfer of oxygen, by means of aerial roots and pneumatophores, is so effective that the mud in the vicinity of underground mangrove roots is less anoxic than that at some distance from the root. Mangroves roots oxygenate their environment (p. 15). Mangroves also have local effects on soil salinity. Figure 4.5 shows the salt concentration along a transect through a belt of mangrove trees and adjoining salt flat in northern Australia. Salinity in the soil under the mangroves is constant, at around 50‰, but immediately outside the mangrove belt it rises sharply. It seems that the presence of mangrove trees prevents extreme hypersalinity. This could be because they remove salt from the soil and get rid of it by salt-gland secretion (see p. 19) or by shedding senescent leaves containing accumulated salt. Alternatively, evaporation (and concentration of residual salt) from the mud surface may be less in the shade of mangrove trees than on open mud. A final possibility is that mangrove soil is permeated by crab burrows (Chapter 6), and water flowing through these may remove high salinity water from the area (Ridd and Samm 1996).

At a smaller scale, water is continuously being taken up through the mangrove roots and, in species such as *Avicennia* and *Aegiceras*, much of the salt

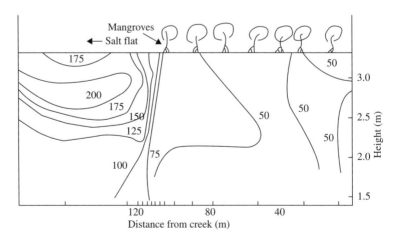

Fig. 4.5 Salt concentration profile for a transect through a mangrove and adjacent mudflat. Salt concentrations are in g/l. Reproduced with permission from Ridd, P.V. and Samm, R. 1996. Profiling groundwater salt concentrations in mangrove swamps and tropical salt flats. *Estuarine, Coastal and Shelf Science* **43**: 627–635, with permission from Elsevier.

is excluded at the root surface. This leads to accumulation of salt just outside the root. This very local increase in salt concentration may reach such proportions as to limit the growth rate of mangrove seedlings (Passioura *et al.* 1992). Given the considerable biomass of a typical mangrove forest, and the demand for nutrients such as nitrate and phosphate (p. 24), it is obvious that mangroves will have an effect on prevailing nutrient levels. Mangrove soil ammonium levels fall by about 75% during the season of active growth (Boto and Wellington 1984).

Mangrove trees transform what might otherwise be relatively uniform mud into a structurally complex and highly productive environment, and affect levels within that environment of oxygen, salt, and inorganic nutrients. The relationship between mangroves and mud is a reciprocal one.

4.2 Seagrass meadows

The distribution patterns of seagrasses, as of mangroves, are the outcome of their response to physical variables, to biotic factors, and to chance. In mangroves the key physical factors are the nature of the sediment, tidal-inundation frequency, and salinity; relevant biotic factors include relations between different species of mangrove tree, as well as with other organisms; and chance factors include the creation of gaps through the random deaths of trees or the irregular impact of hurricanes or typhoons.

Similar factors structure seagrass communities. Tidal fluctuations primarily affect only species that live intertidally, such as *Zostera*, *Halophila*, and *Phyllospadix*. When uncovered by the tide they survive desiccation in much the same way as intertidal macroalgae, principally by lying flat on the mud surface and reducing the surface area for evaporation. The distribution of intertidal species is probably limited in part by competition with algae, and perhaps at the upper shore levels by salt-marsh plants.

Exposure to ultraviolet radiation may also be a limiting factor: ultraviolet (UV) raditation depresses photosynthesis and can damage sensitive cell organelles. The leaves of intertidal seagrasses may show increased levels of UV-absorbing pigments such as flavonoids. A thickened pigment-containing epidermis may absorb up to 99% of UV radiation, protecting the more vulnerable underlying cells. Several intertidal seagrasses, such as *Zostera capricorni*, have thickened leaves. Some intertidal species such as *Halophila ovalis* have thin leaves and may rely on pigment concentration rather than on epidermal thickness. Both *Z. capricorni* and *H. ovalis* also occur intertidally: UV protection may affect the upper limit at which they can survive intertidally, but is clearly not the dominant factor affecting their distribution on the shore (Dawson and Dennison 1996).

Subtidal species are limited by the availability of light for photosynthesis, with different species responding differently to light level. Light penetration sets the maximum depth at which seagrasses can live. Tidal fluctuations in water depth will affect light exposure, but probably rather less than other variables such as turbidity of the water. Even at extreme high tide, mangrove leaves are rarely covered by water, so photosynthesis can proceed at all states of the tide. Where mangroves are light-limited, it is a consequence not of the surrounding medium but of being in the shade of other mangroves.

Soil oxygenation can also affect the competitive interactions between species, and the species composition of a seagrass meadow. Artificially lowering the soil redox potential (p. 8) reduces leaf growth and increases shoot mortality in some species, but not in others, so to some extent natural soil redox potential will affect species composition. For some reason, temperate species seem to be less affected by low soil oxygen than tropical ones (Terrados et al. 1999).

Like mangrove stands, seagrass meadows generally comprise very few species (Figure 4.6). In the Indo-West Pacific region, up to seven species may be found within the same meadow, with *Thalassia hemprichii* usually the most abundant. In contrast, in Western Australia, although comparable numbers of species occur, seagrasses generally segregate into monospecific patches (Duarte 2001). Where an area of habitat is suitable for several of

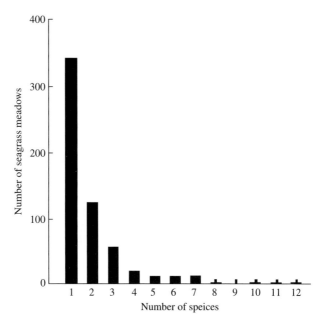

Fig. 4.6 The number of species present in seagrass meadows ($N = 596$). Reprinted from Duarte, C.M. (2001). Seagrasses. In *Encyclopaedia of Biodiversity* (ed. S.A. Levin), volume 5 pp. 255–268. Academic Press, with permission from Elsevier.

the species in the area, but is occupied exclusively by a single species, the explanation probably lies in competition between the two. Less effective competitors are progressively eliminated, culminating in domination by a single climax species. Homogeneity is restored.

Competitive exclusion will occur, of course, only when resources are limiting, which will tend to be the case, sooner or later, in an undisturbed ecosystem. The part played by resource limitation in the outcome of competition has been demonstrated experimentally by artificial manipulation of nutrient supply in seagrass meadows in Florida. Initially, these were homogeneous growths of turtlegrass, *Thalassia testudinum*. The placing of artificial roosts for seabirds was then followed by considerable local nutrient enrichment of the meadow by bird droppings, amounting to 0.68 g of nitrogen and 0.13 g of phosphorus at each roost site. For the first two years the surrounding turtlegrass increased in amount: thereafter it was progressively replaced by *Halodule wrightii*, until after 8 years *Halodule* made up more than 97% of the above-ground biomass. The effect persisted at least 8 years after the nutrients were no longer supplemented by attracting birds to roosts. Where nutrients are limiting, competition leads to the dominance of *Thalassia* and exclusion of other species: when nutrients are not limiting, this is not the case. Seagrasses may also compete with algae (Fourqurean *et al.* 1995; Davis and Fourqurean 2001; Taplin *et al.* 2005).

If a gap appears in a seagrass meadow, then for a time at least, space is no longer a limiting resource. The plants surrounding the space are in a strong position to occupy it by horizontal rhizome extension: the characteristic growth pattern of seagrasses predisposes them to occupy space. But the patch may also be colonized by the random settling of viable rhizome fragments, and by seedlings. Some of these will be of mixed species, depending on the species present in the vicinity, and of the dispersal ability of their propagules: some species are better dispersers, others more effective competitors. The initial infill may therefore comprise species different from the surrounding established adult plants, and until the less successful are eliminated by competition, more species will be present than before the gap appeared. Disturbance and the random appearance of patches will, at least temporarily, promote species diversity. If gaps and other perturbations of a steady state are frequent and severe enough, an ecosystem may never reach a climax state. It is commonly believed that species diversity is higher at intermediate levels of disturbance, with both high and low disturbance levels selecting only those few species that flourish in extreme conditions. In general, seagrass habitats are prone to quite frequent disturbance by such agents as storms, and large grazing herbivores such as dugongs (p. 144); not to mention various forms of human disturbance. Seagrass ecosystems exist mostly in an intensely dynamic—and unstable—equilibrium with their environment, and this instability is certainly one factor determining the composition and structure of seagrass meadows.

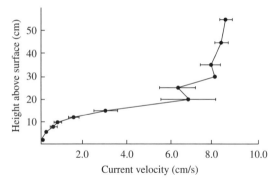

Fig. 4.7 Vertical profile of current velocity within a seagrass meadow of mixed *Halodule wrightii* and *Zostera marina*. Error bars are S.E. From Irlandi, E.A. and Peterson, C.H. (1991). Modification of animal habitat by large plants: mechanisms by which seagrasses influence clam growth. *Oecologia* **87**, 307–318, with kind permission from Springer Science and Business Media.

Not all interactions between seagrass species are competitive. Seagrasses alter their environment in ways that may benefit other species. An established seagrass meadow slows water currents (Figure 4.7), encouraging the precipitation of fine particles, and fine sediments are held together by seagrass rhizomes. Seagrasses, like mangroves, make mud. The reduction in water flow makes it easier for pollen to settle, and for seedlings and rhizome fragments to settle and establish themselves, whereas the rhizomes of an established species provide attachment for the grappling hooks of *Amphibolis* seedlings (p. 48). The canopy of large seagrasses may monopolize light and make it difficult for small species to flourish in their shade; but they may also create a stable habitat in which smaller and more vulnerable species can grow. In Shark Bay, Western Australia, for example, *Amphibolis* and *Posidonia* normally form dense monospecific meadows, but in some areas of sparser cover an understorey of the smaller species *Halodule uninervis* may develop (Walker 1989). Oxygenation of the sediment by leakage from the underground roots of one species may also make conditions more favourable for another.

Although the structure and composition of seagrass meadows gives an impression of uniformity, there are quite complex interactions between species: these are far from being fully understood.

5 The mangrove community: terrestrial components

A mangrove community is more than just an assemblage of trees physiologically adapted to living in brackish water. Living in, on, or around the mangrove trees is a heterogeneous community of organisms, which depend on them for attachment, shelter, or nutrients. The mangrove trees may suffer or benefit from their presence. They may be permanent residents, or may occupy the mangal temporarily, either seasonally or for part of their life cycles. They may occur exclusively in mangrove habitats, or be found in any humid tropical environment. In addition, vast numbers of bacteria and fungi are involved in decomposition of detritus. Although little is known in detail about the species involved, their function within the ecosystem is crucial, and is discussed in Chapter 8. Mangrove-associated animals or plants are of either terrestrial or marine origin. In this chapter we shall consider the terrestrial components of the mangrove community.

5.1 Mangrove-associated plants

Mangrove trees provide a firm substrate on which other plants can grow. Climbers include such plants as the vines *Caesalpina* and *Derris* (Leguminosae), lianes, the mangrove rattan palm (*Calamus erinaceus*) of Malaysia, the orchid *Vanda*, and climbing ferns. These are rooted in the soil, usually on the landward fringe of the mangal, but use mangrove trees as support and may extend their shoots into the more seaward mangrove trees. Some plants are truly epiphytic, and grow entirely on mangrove trees with no direct connection with the soil. These include lichens growing on the bark of trunks and branches, and vascular plants. Among the latter are several species of fern and orchid and, in the New World, bromeliads.

In general, the species of vascular plant associated with mangroves, whether as climbers or true epiphytes, are the same as those that occur in adjacent terrestrial communities. They are unable to tolerate high salt levels, and therefore do not penetrate deeply into the mangrove habitat. There are, however, some apparent exceptions. Some bromeliads, for instance, have succulent leaves and seem to accumulate salt within their tissues. This suggests that they have evolved a degree of salt tolerance in parallel to the mangrove trees on which they grow (Gomez and Winkler 1991).

One group of plants which depend on mangroves for more than physical support are the parasitic mistletoes (Loranthaceae) of Australia and Papua New Guinea, and similar mistletoe species in other parts of the world. Mistletoes have virtually lost their root system and, instead, tap into the xylem of their host with a penetrating structure known as a haustorium. Because the salt concentration in mangrove sap is higher than that of terrestrial trees (Chapter 2), the tissues of the mistletoe must presumably be as salt-tolerant as those of its host. This is probably relevant to the degree of host specificity shown by some mangrove mistletoes. *Amyema thalassium*, for instance, grows only on the mangroves *Avicennia*, *Excoecaria*, and *Bruguiera*. Other species are more indiscriminate.

Another apparent specialization is the close similarity between the leaves of the mistletoe and those of its host. Resemblance between parasite and host is common among parasitic plants. It may be due to mimicry, so that the plant is well concealed among its host's foliage. Many mangrove species are relatively unpalatable to herbivores, and mangrove mistletoes may, by their resemblance, deter potential herbivore attack without the cost of producing their own aversive substances.

Alternatively, the factors that cause mangrove leaves to evolve their particular configuration may apply equally to the mistletoe: an example of convergent evolution, rather than mimicry. Mangrove species themselves, of different taxonomic origins, are markedly similar to each other in leaf structure, implying convergence. Apart from the fact that it uses its host's root system rather than its own, a mistletoe is just another mangrove species and it is not surprising that it resembles its colleagues in appearance.

A further peculiarity of mistletoes is their means of dispersal. In the case of the mangrove mistletoes, this is by the mistletoe bird (*Dicaeum hirundinaceum*). These have a relatively short intestine, and the mistletoe berries appear to have laxative properties. The combined effect of these two features is that berries pass through the bird within 10–45 min, and travel only 3–5 km before being wiped off on branches, where they germinate (Barlow 1966; Hutchings and Saenger 1987).

A final group of epiphytes, more or less confined to mangroves, are the so-called ant-house plants. As their name suggests, these have a special relationship with ants. They are discussed below (p. 78).

5.2 Animals from the land

Apart from the special features discussed in Chapter 2, mangroves are not very different from other, non-mangrove, forests in their vicinity. Indeed, mangroves often shade almost imperceptibly into adjacent habitats. It is not surprising, therefore, that a substantial proportion of the fauna of mangroves comprises species obviously derived from neighbouring terrestrial environments. Most of the major groups of terrestrial animals are represented in mangroves.

5.2.1 Insects

As anyone who has worked in mangroves can testify, insects are generally present in abundance. Mangrove insects include herbivores, feeding on leaves, flowers, seeds, or mangrove propagules; detritivores, eating dead wood or decaying leaves; more general foragers; and predators. Some insects play crucial roles as pollinators (Chapter 2, p. 29). Finally, of course, insects in their turn represent a major food source for predators. Although mosquitoes and ants are often hard to ignore, other insects which do not make their presence felt to quite the same extent may be of even greater ecological significance. Surprisingly little is known of the significance of mangrove insects.

5.2.1.1 *Insect herbivory*

The most obvious signs of insect herbivory are on mangrove leaves, which often show erosion of the leaf margin, holes in the blade of the leaf, or evidence of leaf-miners. Table 5.1 gives estimates of the percentage of leaf area eaten by insects in 14 species of mangrove surveyed at Missionary Bay, Queensland, Australia. Simple measurements of missing leaf area exaggerate the extent of the damage, since holes expand as leaves grow. Allowing for this, the proportion of leaf productivity claimed by insect herbivores in this case is probably, on average, only around 2.1% (Robertson and Duke 1987). Other estimates, from Belize, China, Malaysia, and other parts of the world support the general proposition that insect herbivory typically accounts for only a small proportion—generally under 5%—of leaf production (Farnsworth and Ellison 1991; Lee 1991).

Although insect attack on leaves may on average be low, the extent varies considerably, both within and between species. In Belize, for instance, leaf damage ranged from 4.3 to 25.3% in *Rhizophora*, and from 7.7 to 36.1% in *Avicennia* (Farnsworth and Ellison 1991), while Table 5.1 shows that species vary between 0.3% (*Excoecaria*) and 35% damage (*Heritiera*). Why the variation? The answer lies in considering another question: what limits insect attack? Not all leaves are equally susceptible to attack. *Rhizophora* leaves are quite tough, even leathery, whereas those of *Aegiceras* and

Avicennia are softer and more succulent; and younger leaves are probably physically easier to eat than older ones of the same species.

Leaves vary also in their chemical composition. Plants are not necessarily passive victims of herbivore attack. They produce a variety of substances that may be deterrents to insect attack, actually toxic, or in some other way reduce the value to the would-be attacker. Even salt secretion on mangrove leaves (Chapter 2) may deter some herbivorous insects such as coccids (Clay and Andersen 1996). More importantly, mangroves are often rich in soluble tannins, or various secondary metabolites which inhibit herbivory (Chapter 6). *Rhizophora* leaves have higher tannin levels than *Avicennia*, and lower levels of insect attack. The species with the lowest level of insect damage is *Excoecaria agallocha*. When the leaves of this species are damaged, they produce a toxic latex which strongly irritates human skin and can be lethal to fish. It seems unlikely to be very beneficial to a grazing insect (Robertson 1991). If plants evolve defences against insects, insects in turn can evolve tolerance of defensive chemicals. This is probably why the insect fauna differs between different mangrove species, a result of adaptation by different insect species to different suites of defensive chemicals (Farnsworth and Ellison 1991).

If the potential dangers of herbivory vary, so too do the benefits. Mangrove leaves vary in nutritional value. A simple index of this is the carbon/nitrogen ratio. A leaf with a low carbon/nitrogen ratio is richer in nitrogen and of greater value to a herbivore. It is therefore predictable that insects

Table 5.1 Estimated leaf damage by insects (percentage of leaves with damage, and estimated leaf area removed) for 14 mangrove species at Missionary Bay, North Queensland, Australia. Data from Robertson and Duke (1987) with permission from Blackwell.

Species	Damaged leaves (%)	Area loss (%; ± S.E.)
Acrostichum speciosum	64	3.1 ± 0.3
Aegiceras corniculatum	99	16.0 ± 0.9
Avicennia marina	89	8.8 ± 0.6
Bruguiera gymnorrhiza	51	3.7 ± 0.4
Ceriops tagal	62	6.3 ± 0.4
Excoecaria agallocha	8	0.3 ± 0.7
Heritiera littoralis	100	35.0 ± 1.6
Lumnitzera littorea	73	4.3 ± 0.5
Rhizophora apiculata	63	5.8 ± 0.5
Rhizophora X lamarcki	34	1.4 ± 0.2
Rhizophora stylosa	57	5.1 ± 0.4
Scyphiphora hydrophyllacea	12	0.7 ± 0.1
Xylocarpus australensis	18	3.0 ± 0.5
Xylocarpus granatum	94	10.0 ± 0.7

would eat leaves of *Avicennia*, with a relatively low carbon/nitrogen ratio, in preference to those of *Rhizophora*. Insect herbivory is sometimes higher when leaf nitrogen levels have been raised by enrichment of the soil with bird droppings or artificially, although this has not always been found (Onuf *et al.* 1977; Farnsworth and Ellison 1991; Feller 1995). The extent of herbivory therefore depends on a host of factors affecting the palatability and value of leaves, and these vary with the age of the leaf, probably seasonally, and between species.

Whereas it is relatively easy to assess the extent of leaf damage, it is not always possible to identify the species responsible. Many species are involved: in the mangroves of Belize, more than 66 leaf-eating species have been identified, while the tally of herbivorous insects in the Andaman and Nicobar islands is nearly 200. The most important are probably lepidopteran caterpillars, although the depredations of chrysomelid and lampyrid beetles, and homopteran bugs, may also be significant (Farnsworth and Ellison 1991; Lee 1991; Veenakumari *et al.* 1997).

Occasionally attack by herbivores may reach epidemic proportions. Even *Excoecaria*, with its toxic sap, is not exempt. In one outbreak, in Indonesia, every tree in an area of 5–10 km^2 was completely defoliated by caterpillars of the noctuid moth *Ophiusa* according to the local press 'as if by demons' (Whitten and Damanik 1986).

In the Mai Po marshes of Hong Kong severe defoliation of *Avicennia* by larvae of the pyralid moth *Nophopterix* is an annual event. Figure 5.1 shows the sequence of events. Over about 10 days, *Nophopterix* caterpillars

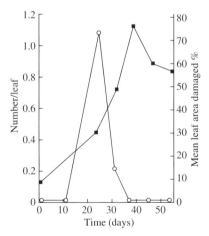

Fig. 5.1 Densities of caterpillars of the pyralid moth *Nophopterix syntaractis* (o) and leaf damage (■) throughout a defoliation event in mangroves of Hong Kong. From Anderson and Lee (1995), with permission from Blackwell.

increase from undetectable levels to a peak of more than one individual per leaf. Twenty days later, leaf damage has risen to nearly 80%. It then declines due to the loss of old, damaged leaves and their replacement by new growth (Anderson and Lee 1995). Such a dramatic event has a drastic impact on the *Avicennia* population. Propagule production is very low, and young seedlings scarce. Very few trees reproduce successfully, except in years when defoliation is less severe than normal. The connection between the presence of caterpillars and failure to produce propagules was confirmed by preventing defoliation by treating the trees with insecticide. In this case flower and propagule production rose to more normal levels.

Defoliating caterpillars may therefore have a significant impact on productivity and reproduction of a mangrove forest. In this case, some 43% of the original leaf biomass was removed, equivalent to 3.4 t/ha. The consequent effect on reproductive rate affects the age structure of the population. Defoliation may, in the long term, threaten the dominance of *Avicennia* at Mai Po, since it will allow other species, whose seedlings thrive in unshaded conditions, to flourish. Finally, much of the leaf material removed is converted to caterpillar faeces and moultskins, and soluble nutrients leach out of this material more readily than from leaf litter. Nutrient cycling, or export, may therefore be facilitated by the caterpillars. Simple leaf-eating may have profound effects on ecosystem structure and function (Anderson and Lee 1995).

Insects also attack developing flowers, fruit, and propagules, both before and after they leave the parent tree. Common culprits are borers such as the scolytid beetle *Coccotrypes*, species of which infest *Rhizophora* in the Caribbean and Australia, and moth larvae. Losses occur throughout the development of buds, flowers, fruit, and mature propagules: in *Avicennia marina* only a tiny minority of flower buds achieve their potential as propagules. Excluding insects does increase fruit survival, but most of this mortality is due to causes other than insect predation (p. 33).

Dispersed propagules frequently show signs of insect damage, which may have been initiated either before or after shedding of the propagule by the parent tree. In one survey, the proportion of damaged propagules ranged from 2.1 to 92.7%, depending on species, location, and year (Table 5.2). Recording visible signs of damage probably underestimates actual losses: not all *Heritiera* seeds, for instance, showed visible signs of insect damage, but when examined more closely, only one out of 435 seeds actually contained an intact embryo. In other species, visible insect damage seemed to have no significant effect on propagule survival or seedling growth (Robertson *et al.* 1990).

Reproductive losses to insects must be seen in the context of massive losses from other causes. After all, if a developing embryo dies because of maternal controls, is eaten by a crab, uprooted by a monkey, or simply fails to find

Table 5.2 Incidence of propagules damaged by insect attack (%) at various sites in tropical Australia. Data from different sites are separated by a comma, data from different years by /. From Robertson et al. (1990), with kind permission from Springer Science and Business Media.

Species	Insect attack (%)
Avicennia marina	59.1, 63.0, 64.8
Bruguiera exaristata	56.0
Bruguiera gymnorrhiza	11.7/44.4, 22.9
Bruguiera parviflora	33.6, 42.1, 48.2, 57.1
Bruguiera sexangula	43.6
Ceriops australis	6.2
Ceriops tagal	7.2, 8.2
Heritiera littoralis	67.8/92.7
Rhizophora apiculata	3.1, 3.5, 6.3, 24.7
Rhizophora stylosa	2.1, 12.5, 33.5
Xylocarpus australoasicus	19.8, 68.4
Xylocarpus granatum	59.1, 75.9

a suitable site in which to take root, the fact that it is also being bored by a beetle will not matter very much. Before assessing the overall importance of insect attack on mangrove reproductive capacity it will be necessary to know a great deal more about its incidence, and about other causes of mortality.

5.2.1.2 Termites

Termites are an important component of the fauna of tropical mangroves. For the most part, they burrow inside the trunks and branches of mangrove trees, particularly towards the landward fringe of the mangrove forest. An exception is a species common in Malaysia (possibly *Nasutitermes*), which constructs an external nest on a tree trunk several metres above the highest point reached by the tide, with narrow galleries snaking down the trunk to the aerial roots, and upwards to the canopy. Termites may be very important in destroying dead wood: a dead tree trunk, if poked, sometimes crumbles and collapses, revealing a mass of termites and little else left of the interior.

5.2.1.3 Ants

Ants are often abundant in mangrove trees, particularly in the canopy, as they are in other tropical forest vegetation. Not much is known about ants in mangrove forests, but their abundance suggests their ecological significance. Some species, such as the leafcutter ant *Atta* of South America, feed directly on mangrove leaves and may do considerable damage. Others are predators, and may benefit mangroves by attacking herbivores. The mangroves *Laguncularia* and *Conocarpus* of Florida may attract ants by means of extrafloral nectaries which secrete a sugary solution, thus paying for their protection.

One conspicuous species is the weaver or tailor ant (*Oecophylla smaragdina*) found in mangroves of the Indo-Pacific. Captain Cook came across these in Australia, where they made a deep impression on him as 'a remarkable kind of ant, that was as green as grass: when the branches were disturbed they came out in great numbers, and punished the offender by a much sharper bite than ever we had felt from the same kind of animal before' (quoted in Simberloff 1983). The species gets its colloquial name from its nests, which are constructed by stitching folded leaves together using sticky secretions from their larvae. *Oecophylla* feeds on the sugary secretions of coccid bugs which suck plant sap, and preys on coccids and other insects.

Some of the subtleties of the interactions between *Oecophylla* and mangroves have recently been elucidated in an elegant series of experiments which experimentally manipulated the ant populations on young *Rhizophora*. *Oecophylla* cannot travel across mud, and move from tree to tree above ground level by contacts between branches. Ants were introduced to previously uncolonized young *Rhizophora* trees by connecting them with string 'walkways'. These trees could be compared with paired trees without ants.

The presence of ants significantly reduced the numbers of herbivorous insects—mainly chrysomelid beetles and tortricid and geometrid moth larvae. Numbers of predatory insects were less affected. Much of this reduction in herbivore density was probably due to the herbivores avoiding leaves marked by deposits of ant pheromones rather than by ant predation. Leaf damage by herbivorous insects was also reduced. Surprisingly, there was less damage by the herbivorous crab *Episesarma versicolor* (p. 113), vastly larger than *Oecophylla* and unlikely to be deterred by it (although Captain Cook might disagree with this). Even more surprisingly, this seemed to apply only to male crabs, not female. This is probably because *Episesarma* males prefer leaves already damaged by insect attack. Reduced insect herbivory means fewer already-damaged leaves. The claws of female *Episesarma* differ from those of males: they can tackle damaged and undamaged leaves indiscriminately and are not affected by reduced insect leaf damage.

Reducing insect and crab herbivory seems not to benefit mangrove trees, at least as far as measurable plant growth is concerned. It appears that insect damage induces compensatory growth in trees; when herbivores are removed this growth is suppressed; and, of course, nest construction takes its toll of mangrove leaves. There are complicated relationships between ants, their mangrove hosts, and other species present, and ants may have both harmful and beneficial effects (Offenberg *et al.* 2004a, 2004b, 2004c).

An even more intimate dependency has been established between ants and certain species of plants epiphytic on mangroves, known as myrmecophytes, or ant-plants. In Australia the *Myrmecodia* plant, which may weigh several kilograms, has a bulbous stem honeycombed with tunnels occupied by the ant *Iridomyrmex* (and, in addition, a butterfly larva). Ants living in such ant-house plants clearly gain protection: is there any advantage to the plant? Another myrmecophyte species, *Hydnophytum formicarium*, has specialized absorptive chambers. Ants deposit their debris here, and it has been demonstrated that when the colony is fed radioactively labelled *Drosophila* larvae radioactive compounds are absorbed into the plant. The relationship is therefore mutual: ants obtain shelter, and the plants a supply of scarce nutrients, particularly nitrogen. Saprophytic fungi growing in the ant galleries probably play a role in releasing soluble nutrients from the ant debris. To make the situation even more complicated, the ants also tend larvae of the butterfly *Hypochrysops*, which feed on the tubers and leaves of the ant plant. An epiphytic plant therefore grows on a mangrove tree, accommodates ants, which tend butterfly larvae and supply nutrients to their host, aided by fungi: this is two plants, one or more fungi, and two animal species interacting (Janzen 1974; Huxley 1978).

Ants are essentially terrestrial animals. It is not surprising, therefore, that the ants found in mangroves are largely arboreal, and construct their nests well above the highest point reached by the tide. Most species are shared with neighbouring terrestrial forest. Some, such as the leafcutter *Atta*, may nest outside the intertidal zone and forage in the mangal only at low tide. A few species are restricted to mangroves. In Australia, *Polyrachis sokolova* is found only on the lower shore, in the *Ceriops* and *Rhizophora* zones. Uniquely, it nests actually in the mangrove mud. Because of the position of the nests, they are inundated in up to 61% of high tides, for periods of up to 3.5 h. Individual *P. sokolova* forage at low tide, returning to the nest before the entrances are covered by the tide. Examination of food remains in the nest shows a broad range of items: small decapod and amphipod crustacea, lepidopteran larvae, and other insects. They have also been observed feeding on bird droppings.

Are these ants adapted to a truly aquatic life? They are able to run or swim on the water surface in order to reach the nest entrance before the tide covers it, but this is not truly aquatic behaviour. The hairy surface of the cuticle can hold an air film during submersion, but ants do not live long under water: at 33°C only 50% survive 3.5 h of immersion. The structure of the nests has been studied by making casts with polyurethane foam. They may be up to 45 cm deep in the mud, and have two entrance holes. When the advancing tide reaches the nest, loose soil collapses into the entrances, blocking them and preventing water from penetrating the passages beneath. In this way ants—provided they get back to the nest in time—survive high tide in air trapped in the nest galleries (Clay and Andersen 1996; Nielsen 1997).

5.2.1.4 *Mosquitoes and other biting insects*

The record for the number of mosquitoes attracted is probably held by William Macnae; over 80 settling on his bared arm within 2 min (Macnae 1968). Many workers in mangroves probably feel they have come close to this. Mosquitoes (and other biting insects, such as the sandflies, or biting midges (Ceratopogonidae)) attract attention because of their nuisance value, and because in many cases they are vectors of diseases such as malaria and yellow fever.

To breed, mosquitoes require shallow and fairly stagnant pools of water. These occur in abundance in mangroves. Rot holes in trees, tidal pools, and the water retained in crab burrows all provide suitable environments for egg-laying and larval development. Different species of mosquito have slightly different requirements. In Australia, *Aëdes amesii* favours rot holes in branches, and *Aëdes vigilax* ephemeral tidal pools in good sunlight to encourage the growth of the phytoplankton on which the larvae feed. Development from egg to adult takes little over a week in the tropics, so a tidal pool left at the height of a spring tide persists long enough for complete larval development. The eggs of *Aëdes alternans* are laid at the edge of pools, and can survive drying out. In East Africa, *Aëdes pembaensis* lays its eggs on the claws of the crab *Sesarma meinerti* and the larvae develop in the crab's burrow. Biting midges do not require standing water for breeding; their larvae are part of the substrate infauna. Mosquito larvae flourish in brackish water. Freshwater mosquito larvae have a group of anal papillae which actively acquire sodium and chloride ions from the surrounding medium. Larvae that live in brackish water, such as those found in mangroves, are surrounded by an abundance of sodium chloride, and do not need special mechanisms to acquire it. Their principal physiological adaptation is that the anal papillae are much reduced in size and, in at least one species, they are impermeable to sodium chloride.

Humans are enthusiastically used as prey, but mosquitoes and midges presumably depend largely on the mammal and bird fauna of mangroves. Although there appear to be no records of mosquitoes feeding on reptiles, this is a possibility. Species of *Aëdes* have been seen feeding on mudskippers, so they are not restricted to warm-blooded targets (Macnae 1968; Hutchings and Saenger 1987).

5.2.1.5 *Synchronously flashing fireflies*

One of the most unforgettable sights in the mangroves of south-east Asia is the synchronized flashing of fireflies. The River Selangor of western peninsular Malaysia is fringed with mangrove trees: as darkness falls, most of these become alive with the lights of thousands of tiny flashes of light. What is most remarkable about these is the regularity of the flashes—about three times a second—and their synchrony. In one tree all of the flashes appear simultaneously, and neighbouring trees are virtually in synchrony.

In Selangor, the flashes come from the firefly *Pteroptyx tener*. Despite their vernacular name, fireflies are in fact beetles of the family Lampyridae. Of the more than 2000 species of this family found throughout the tropics, only a handful synchronize their flashing. The phenomenon appears to occur only in southern Asia and the western Pacific, from east India through Thailand, Malaysia and Indonesia to the Philippines and Papua New Guinea. Different species flash with different frequencies. *Pteroptyx malaccae* of Thailand has a cycle of half a second, varying with temperature: at 25°C the timing is 560 ms, with no individuals more than 20 ms out of synchrony (Figure 5.2). In New Guinea, *Pteroptyx cribellata* has a cycle of 1 s, and other species flash at intervals of up to 3 s.

To achieve synchrony, fireflies must respond to the flashes of their colleagues. Experiments have shown that they can be entrained to the rhythm of a flashing light of appropriate intensity over a range of frequencies, but that the firefly's flash coincides with the stimulating flash only at its 'natural' frequencies. A firefly with a natural frequency of one flash per second, stimulated by a lamp flashing once every 1300 ms, will respond by adjusting its frequency to match that of the lamp—but its flashes precede the lamp signal by 300 ms (or lag by 1 s). In natural circumstances, flashing is presumably regulated by an internal pacemaker, whose frequency is characteristic of the species, entrained by the flashes of other fireflies in the vicinity.

Why flash in synchrony? Clearly it must be some form of display. Only certain trees are chosen, mostly *Sonneratia caseolaris*, and (as a subjective

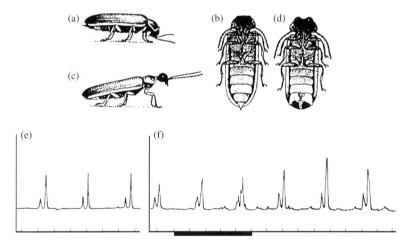

Fig. 5.2 Top: males and females of the firefly *Pteroptyx malaccae*; (a, b) female; (c, d) male. Bottom: the graphs show a recording of the sequence of flashes of (e) an individual *P. malaccae* and (f) a group. The bar on the *x* axis denotes 1 s. Reproduced by permission from Buck, J. and Buck, E. 1976. Synchronous fireflies. *Scientific American* **234/5**: 74–85, with the permission of Nelson H. Prentiss.

impression) those in fairly conspicuous positions. In some areas other mangrove species may be favoured: possibly what makes a tree attractive is the disposition of its foliage, so that a flash can be clearly seen at a distance. Fireflies seem to favour trees which are relatively clear of scale insects and weaver ants: as the two tend to go together it is probably the ants that fireflies are avoiding. A prudent firefly should land in a tree that already has a population of flashing fireflies, advertising that it is ant-free: the numbers will therefore accumulate. Particular trees are consistently used for displays for many years. The displays in one tree in Singapore have been studied for five consecutive years, but reports that Malay fishermen use firefly trees for navigation suggests that individual display trees may be virtually permanent. Individual fireflies probably live for only a few weeks.

Only males flash in synchrony. Although a display tree may have a number of resident males, many more commute in as dusk approaches. Flashing starts soon after sunset, and builds up to a peak within 15–20 min. Females approach, with a flickering light: males may enter a brief flash dialogue with females, or even fly off in chase. Copulation may be accompanied by a dull glow by both sexes, then the female may fly off, flashing rhythmically, to seek a site for egg-laying. It is fairly clear that flashing is used in courtship, but the synchrony of the flashing remains enigmatic. The problem is to work out the individual's advantage in participating in group activity. A plausible hypothesis is that flashing establishes a male's credentials as a potentially acceptable suitor, a male that does not correctly synchronize his flashing being excluded from consideration. Once a group of males has assembled round a female, she then selects the one with, for example, the brightest light. This corresponds to the lek courtship of numerous bird and mammal species. The synchrony of whole trees may merely be a consequence of synchronized activity within small groups within sight of each other (Macnae 1968; Buck and Buck 1976).

5.2.1.6 *Other insects*

Very little is known of other insect groups within mangrove habitats. It seems likely that the insect fauna of mangrove canopies resembles that of terrestrial forests, and that many species are common to both. Numerous butterfly and moth species have been recorded from mangroves. Bees, including the domestic honey bee *Apis mellifera*, feed heavily on *Aegiceras* and *Avicennia* flowers. Many of these species may be important pollinators (Chapter 2). Other species, such as cockroaches, certainly occur in mangroves, but nothing is known of their ecological importance.

5.2.2 Spiders

Little is known of other mangrove invertebrates of terrestrial origin, other than geographically patchy species lists and more or less anecdotal accounts of some of the more conspicuous species. Among other arthropods, the spiders are best known, or perhaps just more conspicuous. Web-building spiders occur, often in abundance. One of the most spectacular of these is the golden silk spider (*Nephila clavipes*) of the New World. Female *Nephila* build an orb web some 2 m in diameter. The female, which occupies the web, may have a body length of about 6 cm. She may also share her web with a number of kleptoparasitic spiders (*Argyrodes*) which build no web of their own, but steal trapped insects from the *Nephila*. Another striking group of web-building spiders belong to the family Gasteracanthidae. These have brightly coloured bodies with projecting spikes which may protect them against predators, and occur in mangroves from the Caribbean to Malaysia and Australia.

As well as the web-building spiders, mangroves often contain numerous wolf and jumping spiders which descend from the trees at low tide and forage over the mud. Their main prey is presumably insects such as ants, but the lycosid *Pardosa* of Malaysia has been seen to catch juvenile fiddler crabs (*Uca*). *Pardosa* is one of the few spiders that seems to be adapted to a semi-aquatic life. Its hairy coat is water-repellent, and the species seems to shelter (and breed) in air-filled burrows in the mud, including abandoned fiddler crab holes (Stafford-Deitsch 1996). With this exception (and a few others) the spiders found in mangrove habitats occur also in neighbouring forest. They are not peculiar to mangroves, nor particularly adapted to the mangrove environment.

5.2.3 Vertebrates

Many vertebrate species of terrestrial origin occur within mangrove habitats, particularly reptiles, birds, and mammals. Other vertebrates, particularly fish, have entered mangroves from a seaward direction. The provenance of some mangrove species is not straightforward: reptiles such as sea snakes and (perhaps) crocodiles, and mammals such as dolphins are more marine than terrestrial in origin, while some mangrove fish may have originated in fresh water rather than the sea. Nevertheless, for convenience, amphibians, reptiles, birds, and mammals are discussed in this chapter, and fish are dealt with in the next among species of marine origin.

5.2.3.1 *Amphibians*

Amphibians are rarely found in brackish or salt water. The only exceptions to this are a few species of frog or toad. One of these is the crab-eating frog *Rana cancrivora*, found throughout south-east Asia, where it is sufficiently abundant to be the most important edible frog of the area. *R. cancrivora* is common in rice fields and ditches, often well inland, so is by no means restricted to brackish water. Both adults and larvae are, however, common in mangrove habitats. Rafts of eggs are laid in small pools of water such as crab holes or drainage channels. Young tadpoles, like those of most frog species, are herbivorous: their gut contents include fragments of vegetation, algae, and mud. Larger tadpoles have a more mixed diet, and adult frogs are of course exclusively carnivorous. The species lives up to its name: in one study of adult frogs from a mangrove area, every single gut examined contained fragments, or even entire individuals, of sesarmid crabs (see Chapter 6). The diet of R. cancrivora from fresh water consists largely of insects (Elliott and Karunakaran 1974).

Given the variety of habitats in which the species occurs, it is clearly euryhaline rather than adapted to a specific salinity range. How does R. cancrivora survive the typical salinities of mangrove habitats? Tadpoles are effective osmoregulators; that is, they maintain their internal osmotic concentration relatively constant over a wide range of external salinities (Figure 5.3). This is probably achieved by the active transport of ions out of the body at high environmental salinity, and by the retention of ions and elimination of water when the surrounding salinity is lower than that of the body fluids. Whatever the physiological mechanisms involved, young tadpoles can survive in concentrations of 40% that of sea water (14‰) and older ones

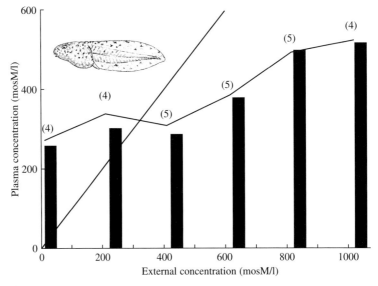

Fig. 5.3 Total osmotic concentration (continuous curve) and concentration of sodium, chloride, and potassium ions combined (bars) in blood plasma of tadpoles of *Rana cancrivora* acclimatized to various external salinities. The diagonal straight line represents the line of equivalence between external and internal concentrations. Concentrations are in milliosmoles/litre; figures in brackets indicate sample size. Reproduced with permission from Gordon, M.S. and Tucker, V.A. 1965. Osmotic regulation in tadpoles of the crab-eating frog (*Rana cancrivora*). *Journal of Experimental Biology* **42**: 437–445. ©Company of Biologists Ltd.

to 50% sea water or higher. At 80% sea water (28‰) approximately 50% of tadpoles still survive, and some still live at greater ion concentrations than that of full sea water (Gordon and Tucker 1965).

Although the evidence is largely anecdotal, metamorphosis of tadpoles into adult frogs may depend on an influx of fresh water at the start of a rainy season. It has been suggested that tadpoles living in brackish water enter a resting state like the condition of diapause in insects, and that further development, and metamorphosis, are triggered by a sharp fall in the salinity of the environment (Macnae 1968). This is unlikely to apply in areas such as Malaysia where heavy rainfall is—to say the least—not an unusual or seasonal event.

Adult frogs have a different, and rather more unusual, method of osmotic regulation. Instead of being osmoregulators and maintaining an imbalance between the osmotic concentration of their internal fluids and that of the exterior, they are partial osmoconformers. Internal osmotic concentration is matched with that of the exterior, at least in a hyperosmotic medium. This is brought about not by manipulating ion levels, but by 'topping up' the ionic concentration of the plasma with the non-ionic solute urea $CO(NH_2)_2$ (Figure 5.4). Other amphibia show slightly raised urea levels in conditions of water shortage, but employing high urea levels to maintain osmotic balance with the environment is a habit shared only with elasmobranch fish (sharks, skates, and rays; Gordon *et al.* 1961). Urea is the standard nitrogenous excretory product of ordinary adult amphibia, and is normally voided at the earliest convenient opportunity. Indeed, given the solubility of urea, it is not an easy substance for a largely aquatic animal to retain, and it is not understood how *R. cancrivora* achieves this. Not only is it soluble and difficult to retain, urea is toxic: the concentrations of urea that occur in the frog's plasma at exterior salinities of 80% sea water, should denature enzymes and affect the binding of oxygen by haemoglobin. Somehow *R. cancrivora* copes with these hazards.

5.2.3.2 *Reptiles*

Compared with amphibians, reptiles are relatively common in mangrove habitats. Reptiles habitually occurring in mangroves include numerous species of snake and lizard, and a few species of crocodile.

Many snake species are not mangrove specialists, but enter mangroves intermittently from adjacent terrestrial habitats only to forage. Pythons, for instance, may enter Australian mangroves in search of roosting flying foxes (see p. 96). A variety of other terrestrial or arboreal snakes also occur, such as the king cobra (*Ophiophagus hannah*) in south-east Asia, and many lesser species. From the opposite direction, sea snakes (Hydrophidae) also utilize the mangrove habitat. The primitive *Laticauda colubrina* breeds on land, but other sea snakes are fully aquatic, even in their reproduction. To some snakes, however, mangroves are the primary habitat. Such specialists include the cat snake *Boiga dendrophila* of Australia and south-east Asia, a poisonous and aggressive species that may reach 1.5 m in length. Its dangerous nature is advertized by the classical warning pattern of a glossy black body ringed with gold bands. Mangrove snakes are common in the Indo-Pacific, but also occur in mangrove habitats elsewhere, an example being the mangrove snake, *Nerodia fasciata*, of Florida.

Snakes are exclusively carnivorous, and their prey in mangroves consists principally of small fish and crabs. On occasions, foraging snakes have been seen to insert their heads into crab or mudskipper burrows to look for prey. Sometimes the roles are reversed. Judging from analysis of stomach contents, juvenile snakes are an important source of food for crabs. A hard exoskeleton may protect crabs against being bitten, and the ability to respire in water gives a crab the advantage in an underwater dispute over which is the predator and which the prey (Voris and Jeffries 1995).

Many lizards are found in mangroves, usually species typical of adjacent terrestrial forests. In south-east Asia and Australia, monitor lizards are

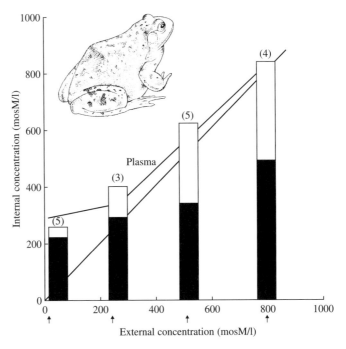

Fig. 5.4 Total osmotic concentration (continuous curve) and concentration of urea (white bars) and sodium, chloride, and potassium ions combined (black bars) in blood plasma of adults of *Rana cancrivora* acclimatized to various external salinities. The diagonal straight line represents the line of equivalence between external and internal concentrations. Concentrations are in milliosmoles/litre; figures in brackets are sample size. Reproduced with permission from Gordon, M.S. *et al.* 1961. Osmotic regulation in the crab-eating frog (*Rana cancrivora*). *Journal of Experimental Biology* **38**: 659–687. ©Company of Biologists Ltd.

among the more striking: the formidable mangrove monitor lizard (*Varanus indicus*) may reach 1 m in length. As with snakes, the lizards of mangrove habitats are all carnivorous.

Even more formidable than monitor lizards are the crocodiles and alligators. In Central America the American crocodile (*Crocodylus acutus*) and the common caiman (*Caiman crocodylus*) may occur in mangroves. In West Africa the nile crocodile (*Crocodylus niloticus*) is found, while from India through south-east Asia and Australia, as far as Fiji, the commonest mangrove species is the estuarine crocodile (*Crocodylus porosus*). Estuarine crocodiles may reach a length of 7 or 8 m. Small crocodiles catch fish and mangrove invertebrates; larger ones also prey on mammals including, should the opportunity arise, humans. The Sarawak Museum in Malaysia recently mounted a display which included, among other memorable items, a wristwatch recovered from the stomach of a crocodile caught in the vicinity a number of years ago.

To live in the mangrove environment, reptiles face the same osmotic problem as other organisms. Figure 5.5 demonstrates the effects of transferring five small American crocodiles between water of different salinities. In sea water there is a loss in mass of 1–2% each day as water is lost to the environment. On transfer to fresh water, the lost water is regained and the animals stabilized at around their original mass. At 25 or 50% sea water the animals appeared to be able to maintain their body mass with no net influx or efflux of water (Dunson 1970). Larger crocodiles are better

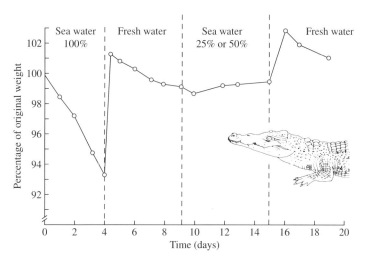

Fig. 5.5 Change in mean body weight of five fasting American crocodiles (*Crocodylus acutus*) exposed to different salinities, as indicated. Reproduced with permission from Dunson, W.A. 1970. Some aspects of electrolyte and water balance in three estuarine reptiles, the diamond back terrapin, American and "salt water" crocodiles. *Comparative Biochemistry and Physiology* **32**: 161–174, with permission from Elsevier.

able to cope with sea water than small: one American crocodile thrived after 5 months in full sea water, and the estuarine crocodile has been found in the open sea at some distance from land.

Mangrove-inhabiting reptiles share some common adaptations that enable them to survive an osmotically hostile saline environment. The keratinized reptile skin is relatively impermeable to water; not completely, but presumably sufficiently so to make it feasible to control water and salt balance by other means. Marine snakes are much less permeable to the influx of sodium ions than freshwater species (Dunson 1978). Many species also have valvular nostrils which further limit water movement.

Some water inevitably enters the body with the food. However, in captive conditions feeding takes place primarily when brackish water is available. Drinking is selective. Adult American and estuarine crocodiles avoid drinking when in salt water, and hatchlings may exploit the surface lens of fresh water that results from rainfall (Mazzotti and Dunson 1984; Stafford-Deitsch 1996). Crocodiles are rather more discriminating than alligators in this respect. Given a choice, they will drink only fresh water. Alligators from freshwater populations are not normally able to discriminate, although animals from estuarine populations of a normally freshwater species did select fresh water (Jackson *et al.* 1996). Mangrove snakes show a similar ability to tell salt from fresh water, in contrast to freshwater snakes which, when placed in sea water, lose water and attempt to compensate by drinking. This makes the problem worse, resulting in catastrophic water loss and death (Dunson 1978).

Even with selective drinking, excess salt inevitably enters the body. The reptilian kidney is unable to excrete urine with a higher salt concentration than that of the blood, although crocodiles (but not alligators) may achieve the same effect by cloacal modification of the composition of their urine (Pidcock *et al.* 1997). Instead of kidneys being the major means of disposing of excess salt, it is eliminated from the body by specialized salt glands. The estuarine crocodile carries between 28 and 40 salt glands, each of which opens by a separate pore on the upper surface of its tongue. Juvenile crocodiles raised in brackish water (20‰) have a blood supply to the salt glands that is three-fold greater than in those reared in fresh water (Franklin and Grigg 1993). Snakes and lizards also have salt glands. In lizards salty water is secreted into the nasal cavity, in sea snakes and file snakes the glands are sublingual, whereas in colubrid snakes such as *Cerberus* they are premaxillary in position.

Reptilian adaptations to a saline environment parallel those shown by the mangrove trees themselves. To a large extent salt water is excluded. If fresh water is available, it is selectively imbibed; but if, despite these precautions, it is necessary to eliminate excess salt from the body, then specialized salt glands are employed.

5.2.3.3 *Birds*

Birds are highly mobile. Many species spend only part of their time in mangroves, either migrating seasonally, commuting daily, or at different states of the tide. They may use mangroves as a feeding area, a nesting area, a refuge from the rising tide, or some combination of these. It is accordingly difficult to analyse their significance to the mangrove ecosystem. This difficulty is compounded by the fact that many accounts of mangrove birds are little more than species lists: detailed investigations of the ecology of mangrove birds are sparse.

Waders probe for buried invertebrates on the mud surface, either among the mangrove trees or on adjacent mudflats. At high tide they either leave the area, or perch on mangrove branches or roots until the surface is again exposed. Herons, egrets, and kingfishers catch fish in the shallow water of mangrove creeks, or prey on mudskippers (see p. 134, Chapter 6) and crabs on the mud surface. Larger fish-eaters, such as storks, pelicans, ospreys, and cormorants, may range further afield and return to the mangroves to roost or breed. The Caroni Swamp of Trinidad is a spectacular example, with densely packed nesting colonies of cattle egrets (*Bubulcus ibis*) and snowy egrets (*Egretta thula*). The scarlet ibis (*Eudocimus ruber*) has been driven by poaching to breed elsewhere, but still roosts in large numbers. The consequence of feeding elsewhere but returning to the mangroves to nest, roost, and deposit large amounts of guano is considerable local enrichment with nitrate and phosphate. Trees used as regular roosts by ibises tend to be taller and to have denser foliage (compare p. 27, Chapter 2; Stafford-Deitsch 1996).

While it is relatively easy to form a general impression of the feeding habits of birds feeding in and around mangroves, it is difficult to quantify their diet, and the relative importance of different items in it. One approach is to catch birds by mist-netting, dosing them with an emetic, and attempting to identify the food remains regurgitated. In one study of the birds of mangrove sites in Panama field observations indicated, unsurprisingly, that three species of egret and six kingfishers foraged on the mud surface in similar ways, and could be regarded as constituting a feeding guild. When individuals were trapped and forced to regurgitate, the diet (based on a total of 46 samples) comprised 45% invertebrate items and 55% fish. The invertebrates included 2.6% gastropod molluscs (snails) and 6.4% crabs: the remainder was made up of spiders, bugs, beetles, and various insect eggs and larvae. Clearly these species, generally categorized as fish-eaters, have a fairly varied diet, and continue foraging on mangrove trees and roots when the tide prevents feeding on the mud.

Many mangrove birds, particularly passerines, depend primarily for food on the trees themselves, or their associated invertebrate fauna. Analysis of the feeding behaviour of the birds of Panamanian mangroves, and their regurgitated food, suggested a number of largely insectivorous guilds. Some 33 species, mainly warblers, gleaned insects from leaves, woodpeckers and similar birds (12 species in all) foraged in the bark of the trees, flycatchers (15 species) caught winged insects in the air, and eight species of hummingbird caught insects and spiders by hovering, as well as feeding on nectar (Lefebvre and Poulin 1997). Analysis of feeding modes in Australian or south-east Asian mangroves indicates similar separation into more or less distinct feeding guilds. One example is shown in Figure 5.6, describing the foraging behaviour of mangrove birds near Darwin, northern Australia.

Species with similar feeding methods are, actually or potentially, competitors. Within a guild, competition is lessened by some divergence in tactics. In the example shown, for instance, the northern fantail (*Rhipidura rufiventris*) hawked for food in tall, well-spaced *Rhizophora* canopy, while the mangrove fantail (*Rhipidura* sp.) operated in dense, low vegetation at the edges of tidal creeks. The three *Gerygone* species, while very similar in foraging techniques, were segregated by plant species, with the mangrove

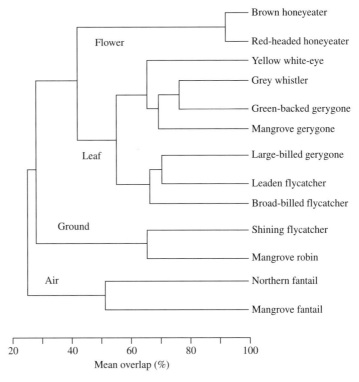

Fig. 5.6 Dendrogram comparing foraging behaviour of 13 species of mangrove birds near Darwin, northern Australia. Reproduced with permission from Noske, R.A. 1996. Abundance, zonation and foraging ecology of birds in mangroves of Darwin Harbour, Northern Territory. *Wildlife Research* **23**: 443–474. ©CSIRO Publishing.

gerygone (*Gerygone laevigaster*) mainly in low thickets of *Avicennia*, the large-billed gerygone (*Gerygone magnirostris*) favouring the landward *Lumnitzera* zone, and the green-backed gerygone (*Gerygone chloronata*) preferring *Ceriops* and *Lumnitzera* (Noske 1996).

Because guild structure is a function of competitive pressures, and of the pool of species in the area, rather than of intrinsic species-specific characters, the composition of guilds differs from site to site. This is true even comparing sites quite close to each other within Australia. Moving further afield, the mangroves of peninsular Malaysia and Singapore show a broadly similar pattern of guilds, composed of completely different species (Noske 1995; Sodhi *et al.* 1997).

How similar are mangroves of Central America and those of south-east Asia? Differences in the methods used preclude any simple comparison: for instance, the Malaysian study produced sufficient data on only 17 species, out of a total of about 125 species found in mangroves in the region, whereas in Panama more than 100 species were covered, including many migrants from North America. Despite these major differences, the overall patterns roughly correspond (Table 5.3). In both cases, the greatest number of species forage on leaves, although the number of species was much greater in Panama. Rather fewer species specialized in foraging on bark. Among the flower specialists, the sunbirds of south-east Asia are counterparts of the hummingbirds of the Americas, and feed similarly on nectar and insects. Relatively more species in Panama than in Malaysia feed on aerial insects.

To most species of birds found in mangroves, the habitat is a convenient feeding resource, but not the only one on which they depend. Long-distance migrants, such as many of the warbler species found in mangroves of south-east Asia and Central America, and the millions of birds of all sorts that annually migrate from central Asia to the Indus delta (see p. 226, Chapter 11), spend only part of their year in mangroves, and the remainder in very different habitats.

Other species may move shorter distances to occupy mangroves on a seasonal basis. In Panama, flooded habitats such as mangroves are buffered from the

Table 5.3 Distribution of insectivorous bird species between the four major foraging modes. Data from Noske (1995) and Lefebvre and Poulin (1997) copyright CSIRO Publishing and Cambridge University Press.

	Malaysia	Panama
Leaves	7	33
Flowers	5	8
Bark	4	12
Aerial	1	15

effects of seasonal rainfall and the resulting variation in availability of insects. Forest-living insectivorous birds therefore move into mangroves when rainfall, and food, are sparse in alternative habitats. In Malaysia, sunbirds (Nectariniidae) move into mangroves when the trees are in flower. As their family name suggests, they feed on nectar, as well as on small insects. When flowering ceases, the birds move out into adjacent rainforest. Sunbirds are the principal pollinators of several species of *Bruguiera*. In Australia they are replaced by the similar honeyeaters (Meliphagidae), and in Central America by hummingbirds (Trochilidae; Noske 1995).

A few species are more or less totally dependent on mangroves, and are rarely found elsewhere. In Australia, for instance, more than 200 species of bird have been recorded from mangroves. Of these, only 14 species of passerine are virtually confined to mangroves, with a further 11 species found predominantly in mangroves, but frequently occurring elsewhere. A similar pattern obtains elsewhere: in New Guinea with 10 species, and Malaysia with around eight species that are regarded as mangrove specialists. Even these figures may be misleading: of the apparent mangrove specialists in Malaysia, four occupy different habitats in other parts of the world, one example being the cosmopolitan great tit (*Parus major*), which is distributed from western Europe to China. Elsewhere in the world, even fewer species are restricted to mangroves: in Africa, there is just a single species of sunbird, and in Trinidad and Surinam a single species of crab-eating hawk (Noske 1996).

Why do so few birds specialize in the mangrove habitat? Africa and Central America have fewer species of mangrove tree and, smaller areas of mangrove, than Australia or south-east Asia. Either factor might contribute to the paucity of mangrove specialists. However, even in the extensive mangroves of south-east Asia and New Guinea relatively few of the bird species found in mangroves are mangrove specialists. Why is this? One reason might be that there is little scope for niche specialization, compared with terrestrial forests. The physical structure of a mangrove environment is simpler than that of most tropical rainforests, with virtually no understorey, less layering of the canopy, and a degree of convergence in leaf size and shape between the different tree species (p. 23, Chapter 2). In addition, regular inundation of the forest floor eliminates the possibility of ground-living specialization. Adaptation, as generalists, to a relatively simple forest structure may be the principal reason why mangrove birds seem to be better than rainforest ones at colonizing human-made habitats. Villages, plantations, urban parks, and gardens share with mangroves a relatively open structure with few vegetation layers and relatively few tree species (Noske 1995).

A second possibility is that mangroves are typically unstable habitats, characterized by accretion and erosion, and the rapid colonization of canopy gaps (p. 61, Chapter 4). Mangrove birds may tend, as a result, to be adaptable and capable of rapidly exploiting new opportunities, rather than narrowly specialist and restricted in their requirements.

Finally, the interchange of species between mangroves and other habitats, such as rainforest, may have limited the emergence of mangrove specialists. It is noticeable that there are proportionately fewer mangrove specialists in New Guinea, where rainforest usually abuts mangrove forest, than in Australia, where this is less common. Within Australia, there are fewer specialists in the mangroves of Queensland, which are extensive and contiguous with rainforest, than in north-western Australia where this is not the case. Mangrove specialists are more likely to arise where there is geographical isolation to prevent gene flow between an incipient specialist and its parent species, and where there is less exposure to rainforest competitors or invading migrants. Conversely, some species may have become restricted to mangroves by being squeezed out of rainforest by more effective competitors (Ford 1982).

Looking at the ecology of the mangrove ecosystem as a whole, it makes little difference whether a particular resident bird species occurs also in other habitats: abundance, diet, and total food intake are more important considerations. Although there is little reliable data on abundance, population densities of mangrove birds appear in general to be low compared with, say, rainforest. This may be related to the relatively simple forest structure, tidal restriction of feeding time, and possibly limited nesting sites. Migrant and commuter species have more unpredictable effects. A species that feeds elsewhere, and resorts to mangroves to roost or breed, is likely to provide a net input of nitrates and phosphates. In contrast, flocks of seasonal migrants which use mangroves to replenish energy reserves for a long passage may represent a significant export of carbon. Even less is known—much less quantified—about the role of transient bird populations than about that of residents.

5.2.3.4 *Mammals*

As with birds, few mammals are characteristic of mangrove habitats. Most species found in mangroves are those that occur in adjacent habitats, although in some cases mangroves harbour species that have been eliminated elsewhere. Those that do occur in mangroves avoid immersion by being either highly mobile or tree-living. The only completely aquatic mammals found in mangrove creeks and rivers are dugongs and cetaceans. Most coastal mangroves are periodically visited by dolphins and porpoises. In the great Sundarban mangroves, lying between India and Bangladesh, there are also the rare freshwater Ganges and Irrawaddy dolphins (*Platanista gangetica* and *Orcaella brevirostris*). Otters, also fish-eaters, are common in south-east Asian mangroves. Other small carnivores may visit mangroves in search of fish or crabs. These include the fish cat (*Felis viverrima*) and mongooses (*Herpestes* spp.) in Asia, bandicoots (*Perameles* spp.

and *Isoodon* spp.) in Australia, and the raccoon (*Procyon cancrivorus*) in Central America.

Terrestrial herbivores forage for mangrove seeds or foliage at low tide. Examples in South American mangroves are the agouti (*Dasyprocta*) and the key deer (*Odocoileus*). In other parts of the world antelopes, deer, wild pigs, and rodents may be common. In the Sundarbans pigs and the axis deer (*Axis axis*) constitute the main prey of the Bengal tiger (*Panthera tigris*): until relatively recently the Javan rhinoceros (*Rhinoceros sundaicus*), water buffalo (*Bubalus*), and swamp deer (*C. duvauceli*) were also common. Domestic camels and buffalo are major eaters of foliage in the mangroves of Arabia (Figure 5.7) and in the Indus delta of Pakistan (Chapter 11).

None of these species is a mangrove specialist. Their association with mangroves is either temporary, or—as with the Bengal tiger—a result of their elimination by humans from alternative terrestrial habitats. Very few mammals are exclusive to mangroves. The Australian rat *Xeromys myoides* forages for crabs at low tide among *Avicennia* and *Rhizophora*, and builds a nest of leaves and mud among the buttress roots of *Bruguiera*, above the level of neap high tides. Although its fur is water-repellent, it has not been seen to swim, so probably avoids high tides by climbing trees (Hutchings and Saenger 1987). This species is rare: even rarer is Cabrera's hutia (*Mesocapromys angelcabrerai*), found in permanently immersed *Rhizophora* on a group of small islets of southern Cuba, and known only from a few specimens collected in the mid-1970s.

Fig. 5.7 Camels feeding on mangrove foliage in the Arabian Gulf. Photo by A.R. Dawson Shepherd.

Monkeys are often common in mangroves. Some are omnivores: in southern Senegal (West Africa) the vervet monkey (*Cercopithecus* sp.) eats both fiddler crabs (*Uca tangeri*) and *Rhizophora* flowers, fruit, and young leaves, while in south-east Asia macaques (*Macaca* sp.) forage on the mud for crabs and bivalve molluscs. They are also in the habit of uprooting *Rhizophora* propagules. Few of these are even slightly damaged, much less eaten, so the motivation is not clear. Whatever the reason, monkey damage is a major problem for mangrove-replanting projects (see Chapter 11), and may also be a significant factor in natural regeneration and community structure of mangrove ecosystems (p. 95).

Some monkey species are more exclusively herbivorous. In south-east Asia, these include colobine monkeys such as langurs or leaf monkeys (*Presbytis* spp.) and the striking Proboscis monkey (*Nasalis larvatus*). Although their diet includes fruit, colobines are specialized in eating leaves. They have a specialized stomach divided (like that of ruminants) into a number of chambers. The forechambers contain anaerobic bacteria, capable of digesting cellulose and dealing with the anti-feeding substances produced by plants, which deter many herbivores. Mangrove leaves are particularly rich in tannins and similar feeding deterrents (p. 107, Chapter 6). *Sonneratia* appears to be the preferred mangrove tree, where available, followed by *Avicennia* (Salter *et al.* 1985).

Proboscis monkeys are exclusively found in Borneo and restricted to mangroves and riverine forests. They form small social groups, typically with a single male. The male has the prominent pendulous nose—which shoots out into a horizontal position when the male honks to deter intruders—that gives the species its common, and scientific, names. It has been estimated (in an inland riverine forest, rather than a mangrove area) that the average density of proboscis monkey groups is 5.2/km^2. Taking into account the mean body mass of the monkeys, this works out at an average monkey biomass per square kilometre of just under 500 kg. Experiments on captive monkeys suggest a daily food intake of around 12% of body weight, explaining their pot-bellied appearance and almost continual feeding. This amounts to 60 kg of leaves and fruit/kg^2 each day, or more than 21 t annually: this is an impressive total, and an indication of the potential ecological significance of the species (Yeager 1989; Dierenfeld *et al.* 1992).

The abundance of insects in mangroves attracts large numbers of insectivorous bats (Microchiroptera), and a mangrove forest may be exploited by numerous species. Competition results in species foraging in different microhabitats within the mangal. Table 5.4, for instance, shows how the aerial environment was partitioned between the bat species of the mangroves of Kimberley, Western Australia. Even species apparently foraging in the same zone showed differences in technique: when measures of wing

Table 5.4 Time allocation of insectivorous bats foraging in Kimberley mangroves, based on position of bat when observed or captured. The figures represent percentages of the total observations (N) made on each species. OC, more than 4 m above canopy; AC, just above (<4 m) canopy; BS/O, beside stand in the open; BS/A, beside stand but against surfaces; IS, inside stand. From McKenzie and Rolfe (1986) reproduced with permission from Blackwell.

Species	Percentage of total observations made on each species					N
	OC	AC	BS/O	BS/A	IS	
Taphozous flaviventris	88	10	2			42
Taphozous georgianus	90	10				20
Chaerephon jobensis	82	16	2			44
Mormopterus loriae	14	55	32			22
Pipistrellus tenuis			22	67	11	36
Chalinolobus gouldii		8	69	15	8	13
Nycticeius greyi		2	24	52	22	46
Nyctophilus arnhemensis				14	86	51

morphology—hence of flight pattern—were taken into account, there was virtually no overlap between species (McKenzie and Rolfe 1986). The impact of insectivorous bats on the insect population must be considerable. Judging from estimates for temperate species, a single bat may consume between one-quarter and one-third of its body weight in insects in a single night. A bat weighing 30 g might therefore eat 10 g of insects, amounting to perhaps 5000 individuals, each night.

Fruit bats and flying foxes (Megachiroptera), exclusively Old World, may also be abundant in mangroves, using the trees as a roost as well as a source of food. Roosts of flying foxes (*Pteropus*) in Australia have been estimated to comprise as many as 220 000 individuals (Macnae 1968). The location, as well as the vast numbers, protect against many potential predators. Most fruit bats feed on nectar as well as on fruit, and it is this which attracts many species to feed in mangroves. In peninsular Malaysia, two of the more important species are the long-tongued fruit bat *Macroglossus minimus*, and the cave fruit bat *Eonycteris spelaea*, which eat *Sonneratia* nectar and pollen. Long, projecting stamens deposit large amounts of pollen on the fur of the bat: the species is a major pollinator of *Sonneratia*. Possibly because of the wear and tear resulting from each visit by a bat, the flowers of *Sonneratia* last for only a single night, and each morning the mud beneath a *Sonneratia* stand is covered by shed stamens.

Macroglossus has a very close association with *Sonneratia*, and in western Malaysia has never been recorded away from mangrove areas. Specialization in *Sonneratia* is possible because, at least in one Malaysian site studied, *Sonneratia caseolaris* is available as a food source throughout the year, while two other species, *Sonneratia alba* and *Sonneratia ovata*, flower in flushes and collectively are available for about three-quarters of the year. *Eonycteris* has a different strategy. It roosts well inland. One massive *Eonycteris* roost is in the limestone caves of Batu, not far from Kuala Lumpur. Tens of thousands of bats may be present. Analysis of pollen in guano from these caves showed that *Sonneratia* was one of the most frequented species. Batu is 38 km, as the bat flies, from the nearest mangroves, so the nightly foraging range is considerable. Many other species of plant are visited. One of these is the commercially important durian (*Durio zibethinus*), a fruit described by some as delectable, and by others as redolent of sewage. *Eonycteris* is a major pollinator. Durians flower and fruit for only limited periods, and pollinating bats therefore require alternative sources of food. Were it not for the availability of *Sonneratia* as a stopgap, durians would not be pollinated and the crop would fail (Start and Marshall 1976).

Both insectivorous and nectar-eating bats cover large distances when feeding, and neither depends exclusively on mangroves. For this reason, they represent important links between mangroves and terrestrial habitats: the mangrove habitat cannot be viewed in isolation from its surroundings.

6 The mangrove community: marine components

Mangroves offer opportunities to marine organisms, providing both a physical environment and a source of nutrients. Pneumatophores and prop roots greatly expand the surface area available, and provide a hard substrate, in contrast to the surrounding mud, while the primary production of mangroves supplies an energy source to many organisms.

6.1 Algae

Apart from the hard substrate offered by mangrove pneumatophores, aerial roots, and trunks, the mangrove soil also provides a surface on which photosynthetic algae can grow. Mostly, these are unicellular diatoms. Photosynthetic blue-green Cyanobacteria are also present on almost any available surface. (Cyanobacteria are often erroneously called blue-green algae: algae are eukaryotes and Cyanobacteria are prokaryotes.) In addition to their role as primary producers, microscopic algae also alter the texture of the soil by affecting particle size and binding soil particles together with their mucous secretions, while some blue-green Cyanobacteria fix atmospheric nitrogen (see also Chapter 2).

Macroalgae are rarer on the mud surface, as most species require a firm surface on which to attach. There may, however, be permanent patches of unattached red algae (Rhodophyta) such as *Gracilaria* and *Hormosira*. *Gracilaria* is extensively cultivated for the extraction of agar, and the natural populations found in some Malaysian mangroves are so abundant that their commercial exploitation has been considered. *Hormosira* is sometimes so dense on mudflats adjacent to mangroves that it reduces the chances of successful settlement by *Avicennia* seedlings by as much as by 80%, thus inhibiting the spread of mangroves into otherwise potentially suitable areas (Clarke and Myerscough 1993).

Pneumatophores and aerial roots are colonized by a film of diatoms and unicellular algae, as well as by a turf of small red algae. Most characteristic is a community of *Bostrychia*, *Caloglossa*, *Murrayella*, and *Catanella* which, in various permutations of species, is found virtually throughout the tropics. The assemblage is often termed a bostrychietum, after its principal constituent. These species seem tolerant of a wide range of salinity conditions and *Bostrychia*, at least, is highly resistant to desiccation. All three genera grow better in shady conditions. In sunlit patches the dominant species are green algae (Chlorophyta) such as *Enteromorpha*.

An accumulation of algae probably impedes the efficiency of aerial roots and pneumatophores by blocking lenticels, while algal growth decreases the survival rates of *Rhizophora* seedlings, and algae must be periodically removed to ensure the success of replanting programmes (Chapter 11).

6.2 Fauna of mangrove roots

Mangrove roots host a range of animal epibionts. The distribution of these often shows a pattern of zonation like that of other hard-bottomed shores. Barnacles (*Balanus* spp.) are among the most conspicuous fouling organisms on the aerial roots of *Rhizophora* and the pneumatophores of *Avicennia*. They may even settle on leaves rather than on the trunk, where these are submerged at higher tides: a short-sighted policy, given the relatively rapid turnover of leaves compared with the trunk surface. Barnacles are filter-feeders, depending on mangroves only for a stable anchorage. Barnacle encrustation can reduce root growth. When their natural predators were excluded from the aerial roots of the red mangrove, *Rhizophora mangle*, in Costa Rica, *Balanus* coverage increased dramatically, reaching more than 80%, blocking lenticels and reducing gas exchange and respiration: root production fell by 52%. The key predators in this case were the snails *Thais kiosquiformis* and *Morula lugubris*, and the hermit crab *Clibanarius panamensis*. These species have a critical role in maintaining mangrove productivity (Perry 1988).

As well as encrusting organisms such as barnacles, mangrove roots are also afflicted with animals which burrow into the root tissue. These include molluscs of the families Teredinidae (shipworms) and Pholadidae (piddocks), and the isopod crustacean *Sphaeroma*. Teredinids are particularly important in destroying dead wood (pp. 130, 158). As with many other herbivorous animals, they cannot themselves digest wood. Like termites, they contain symbiotic cellulose-digesting and nitrogen-fixing bacteria which carry out the difficult biochemistry of dealing with the more recalcitrant plant materials.

Wood-borers can be highly damaging. The aerial portions of roots are particularly badly affected by *Sphaeroma*, probably because closer to the mud surface predators prevent larvae from settling. Isopod boring can dramatically reduce root production and, in serious infestations, even kill roots and cause the collapse of mangrove trees (Svavarsson *et al.* 2002). On the other hand, *Sphaeroma* may benefit mangroves. Damage to the growing root tip can cause branching, so that the number of roots actually reaching the soil surface is increased. Enhanced anchorage will make a tree more stable (Simberloff *et al.* 1978).

The subtidal portions of *Rhizophora* roots usually carry a much richer growth of sessile epibionts, animal and plant. As well as barnacles, these can include oysters (*Crassostrea*) and other bivalve molluscs, sponges, tunicates, serpulid and sabellid annelid worms, hydroids, and bryozoans. These in turn attract a variety of mobile browsing or predatory animals. There is no evidence of adverse effects of subtidal fouling organisms: indeed, species of encrusting sponge have been shown to benefit their host by supplying it with soluble nitrogenous nutrients (p. 25).

If mangrove root-fouling communities are considered in terms of functional groups, they appear relatively consistent and predictable. The dominant organisms are those which filter-feed close to the surface, such as barnacles, ascidians, bryozoa, tube-building polychaete worms, and bivalve molluscs. There are a number of predators, such as thaidid molluscs, and some more general foragers, including species of crab. These broad categories, however, contain many species and a great deal of variation in space and time. The richness and diversity of the assemblages of mangrove root fauna, and the considerable variation between sites, and even between neighbouring roots, raise questions of what factors determine community structure and the species present.

Broad biogeographical considerations have an influence. Mangrove root communities of the tropical western Atlantic are dominated by sponges and tunicates, while those of similar latitudes in the eastern Pacific have abundant barnacles and low coverage of sponges and tunicates. In the Indian Ocean and west Pacific oysters are often more abundant. In total species numbers, the Caribbean communities are richest, with more than 100 species of invertebrate and algal epibionts recorded. This is surprising, since the Indo-West Pacific, as a biogeographical region, is in general more species-rich. The conclusion could, of course, be due simply to greater effort put into investigating root communities in the Caribbean than elsewhere.

The processes that structure a community are complex and varied, and different species respond differently, and at different scales of time and space. Species distributions are affected by gradients in physical variables: at the vast scales of climate and latitude, at the vertical scales of a tidal range, or at even smaller scales. Communities may show successional changes, culminating in a relatively stable and predictable community as rapid colonizers are replaced progressively by more effective competitors. Periodic disturbances, such as storms, may frustrate this process and maintain a balance between colonizers and competitors. Smaller-scale and probably more frequent physical disturbances, or the action of grazers or predators, can create gaps, filled sporadically by whatever colonizing species are available. Selective predation, of course, can alter the balance between prey species. Or the composition of a community can be dominated by chance and the vagaries of the supply of larvae in the plankton: most encrusting marine animals have a planktonic larval stage. Those that do not, or whose larvae disperse only short distances, can dominate areas for no reason other than proximity of a parent.

Which of these general considerations are most important to mangrove root-fouling communities? The most illuminating analyses of community structure come from studies of mangroves in the western Atlantic region, particularly in the Belize cays. Variation in community structure on *Rhizophora* roots was investigated over spatial scales ranging from around 1 cm (between front and back of the same root) to 10 km comparisons between cays), and over a period of more than a year. The results are complex, but the general conclusion is that the variable supply of larvae is important in determining the structure of the epibiont community over short time scales, and over small and very large spatial scales. Over long time scales, and at intermediate spatial scales, variation in physical factors may be more important (Farnsworth and Ellison 1996).

What applies to the mangrove root communities of the Belize cays may not, of course, apply in other areas. In the Florida Keys, mangrove root-fouling communities vary a great deal. Much of the variation is probably due to physical disturbance, particularly from strong tidal flows (Bingham and Young 1995). In contrast, root epifaunal communities on mangroves in Venezuela (again, *Rhizophora*) are highly stable in abundance and species composition. This is possibly because, unlike Florida, the tropics are virtually aseasonal, and factors like the supply of colonizing larvae therefore do not fluctuate greatly (Sutherland 1980).

6.3 Invertebrates

Animals of the root-fouling community use mangroves primarily as a hard substrate, a suitable site for attachment. Many invertebrates depend on mangrove productivity either directly or indirectly. Some species eat shed leaves or reproductive products. Others ingest fine organic particles, either suspended in the water column, as filter feeders, or from the settled mud.

Others again are predators or general scavengers. Mangrove invertebrates include representatives of many phyla, including molluscs, arthropods, sipunculans, and nematode, nemertean, platyhelminth, and annelid worms. The most abundant and conspicuous, however, are crustaceans and molluscs.

6.3.1 Crustaceans

Among the most abundant and diverse of mangrove crustacea are the Brachyura, or true crabs, and among the Brachyura the dominant mangrove species belong to two families, the Grapsidae and Ocypodidae. The former is now generally regarded as a superfamily Grapsoidea, comprising a number of families formerly regarded as subfamilies: the most important of these in mangroves is the family Sesarmidae (formerly subfamily Sesarminae). The Ocypodidae include the fiddler crabs (*Uca* spp.). Other Crustacea commonly associated with mangroves include predatory swimming crabs of the family Portunidae, hermit crabs, the commercially important penaeid shrimps (Chapters 9 and 11), a variety of burrowing species such as the mud lobster *Thalassina*, amphipods, and isopods.

6.3.1.1 *Crabs*

Grapsoid crabs include numerous species, worldwide: many live in brackish, or even fresh water; some are amphibious, and a few virtually terrestrial. It is not surprising, therefore, to find that many grapsoids are characteristic of mangrove habitats, where they forage at low tide on the exposed mud, or on mangrove pneumatophores or tree trunks. At high tide they avoid predators such as fish by retreating into burrows in the mud. Some are more active by day, others by night.

Among the grapsoid crabs, the most important, in number of species and abundance, belong to the Sesarmidae, particularly to the genus *Sesarma* and closely related genera. These are found throughout the tropical and subtropical areas of the Atlantic, Indian and Pacific Oceans, particularly in mangrove habitats. Densities of 50–70 animals/m^2 (equivalent to a biomass of up to 20 g/m^2) are not unusual. Other grapsoid crabs are also characteristic of mangroves (e.g. *Goniopsis* species in the Caribbean), whereas some, such as *Metopograpsus*, occur frequently in mangroves but are also common in other habitats (Figure 6.1).

Sesarmids closely resemble each other. The most distinctive features are a trapezoidal-shaped carapace, not heavily calcified, typically showing a pattern of fine lines of short setae. Ventrally, the carapace plates that lie towards the front of the body, below the orbits, have a well-marked regular grid-like pattern of short bristles. This structure is important in respiration and, probably, in temperature control (see p. 122). The carapace is often cryptic in colour, matching the colour of the mud, although the chelae may be brightly coloured, presumably reflecting a signalling function. When

Fig. 6.1 A grapsid crab (*Metopograpsus* sp.) foraging on an *Avicennia* pneumatophore. Photograph: Hunting Aquatic Resources.

disturbed, sesarmids generally show considerably more agility than biologists. Because of this, and because they (and the biologist attempting to catch them) are frequently covered in mud, they are not the most amenable of animals to identify in the field. Some sesarmids are shown in Figure 6.2. Because of the similarity in morphology, sesarmid taxonomy is—to say the least—complex and confusing, with many species inadequately described, not formally described at all, or described more than once under different names. This situation is one that mangrove biologists learn to live with.

Leaf eating by crabs

Grapsid crabs are often fairly general feeders, depending mainly on scavenging and opportunistic predation. The family as a whole, however, shows a propensity for herbivory. This is particularly true of *Sesarma* and its relatives. Herbivory in sesarmids includes the scraping of epiphytic algae from the surface of mangrove roots. The most important form of crab herbivory, however, is the eating of leaves and the reproductive products of mangrove trees. It is of little direct importance to the tree what happens to fallen leaves, and to the discarded components of flowers. Removal of live and photosynthetically active leaves does happen, but is relatively rare, while the destruction of propagules is frequent and highly significant.

Of the various contributions mangrove trees inadvertently make to herbivorous crabs, fallen leaves are probably quantitatively the greatest. As a result, leaf eating has been studied extensively. The most important herbivorous crabs are undoubtedly sesarmids, but other species must also

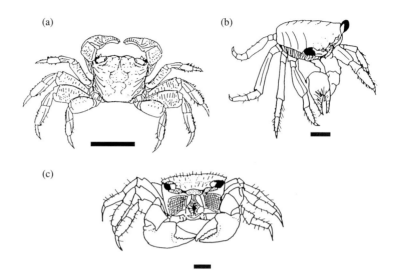

Fig. 6.2 Species of sesarmid crab: (a) *Parasesarma plicata*, (b) *Aratus pisonii*, and (c) *Neosarmatium smithi*. *Neosarmatium* shows the grid-like structure of the epimeral plates, involved in the recirculation and oxygenation of water from the gill chambers. Scale bars, 1 cm. From Jones, D.A. 1984. Crabs of the mangal ecosystem. In *Hydrobiology of the Mangal* (Por, F.D. and Dor, I., eds), pp. 89–109, Figure 1 (part). Dr W. Junk, with kind permission from Springer Science and Business Media.

be considered. The Indian Ocean land crabs *Cardisoma* and *Gecarcoidea* (family Gecarcinidae), and the Central American hairy land crab *Ucides* (family Ocypodidae) are locally important.

Crabs may manoeuvre entire leaves into their burrow, or eat leaves or leaf debris on the mud surface. Food is ingested by manipulating and positioning it with the claws and external mouthparts. It is then broken into smaller portions with the mandibles and passed into the stomach. A crab stomach contains an array of ridges and hooked teeth, which act against each other and grind the food into smaller particles, which are then passed back down the gut for digestion and absorption (Figure 6.3). Some crabs secrete enzymes such as cellulase: this has not been demonstrated in sesarmids, but would certainly be useful in breaking down plant material.

Analysis of stomach contents shows the importance of mangrove leaves to the diet. In many cases the stomach seems to be stuffed full of leaf fragments. In terms of volume, very few other constituents of the diet appear significant—although, of course an item minor in volume may be of major significance if it supplies a requirement (such as nitrogen) in which plant material is deficient. Typical estimates in some representative species indicate that more than 90% of the foregut contents may consist of mangrove leaf fragments (Table 6.1). The diet of sesarmids is dominated by mangrove leaves.

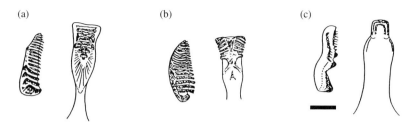

Fig. 6.3 Gastric teeth of (a) the herbivorous sesarmid *Aratus pisonii*, (b) a deposit-feeding fiddler crab (*Uca* sp.), and (c), for comparison, a carnivorous portunid swimming crab *Callinectes sapidus*. The dorsal and left lateral tooth are shown in each case. Scale bar, 1 mm. From Warner (1977).

Table 6.1 Principal components of the diet of representative sesarmid crabs, given as estimated percentage by volume of the foregut contents. Data from Leh and Sasekumar (1985), and the author.

	Dietary components (estimated percentage by volume of foregut contents)				
	Higher plant	Other plant	Animal	Sand, mineral	Other, unidentified
Perisesarma eumolpe	91	1	4	4	0
Perisesarma onychophora	83	1	13	2	1
Parasesarma plicata	67	1	1	5	27
Episesarma versicolor	90	0	7	3	0
Sesarmoides kraussi	93	0	5	2	0
Cleistocoeloma merguiensis	97	0	2	1	0

Direct measurement of feeding rates shows that South African *Neosarmatium meinerti*—a relatively large species—can eat up to 0.65 g of dry weight/day of mangrove leaf material, equivalent to an energy intake of approximately 13.0 kJ/day. Taking into account the weight of the crab, typical feeding rates may be around 2% of the crab's weight per day (Ólafsson *et al.* 2002). Ingestion does not necessarily mean that mangrove material is actually assimilated, since it might pass through the gut without being absorbed. Assimilation efficiencies of mangrove leaves by sesarmids, tested experimentally, are low: from 9 to 32%, depending on the age and state of the leaves. Leaves that had spent some time on the mud were more readily assimilated than those removed directly from trees (Kwok and Lee 1995).

Extrapolating from laboratory experiments is risky. The crabs had little choice but to eat the leaves offered to them, and the absorption efficiencies measured might not reflect their dependence on mangrove material in natural circumstances. The importance of mangrove material in the natural diet of sesarmid crabs can be assessed by investigating the the relative proportions of different stable isotopes of carbon in mangrove leaves, and

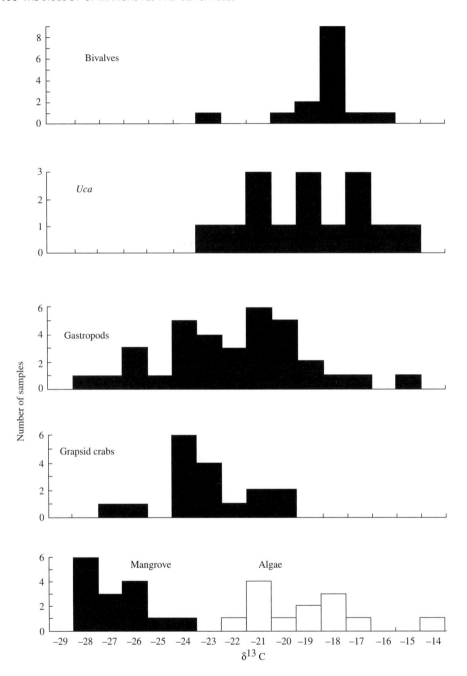

Fig. 6.4 Distribution of $\delta^{13}C$ values of samples of mangrove and algal material, and in the tissues of animals collected from within mangrove swamps, coastal inlets, and offshore. From data in Rodelli, M.R., Gearing, J.N., Gearing, P.J. et al. 1984. Stable isotope ratio as a tracer of mangrove carbon in Malaysian ecosystems. *Oecologia* **61**: 326–333, with kind permission from Springer Science and Business Media.

in the tissues of crabs thought to feed on them. The carbon fixed by mangrove leaves during photosynthesis is predominantly the most abundant isotope, ^{12}C. A small proportion is ^{13}C; the exact proportion varies between mangrove species, but mangroves as a group have a characteristic $^{13}C/^{12}C$ ratio in their tissues which differs from the ratio found in, for example, seagrasses or algae. The ratio, relative to a standard, is expressed as a $\delta^{13}C$ value, in parts per thousand. Mangrove leaves have $\delta^{13}C$ values in the range -24 to $-30‰$. When organic matter is assimilated by an animal, there is some preferential absorption of ^{13}C rather than ^{12}C, but this effect is small and raises the $\delta^{13}C$ value (makes it less negative) by perhaps only 1–2‰ at each trophic level. This means that the $\delta^{13}C$ value of an animal's tissues is a good indicator of the value in its food.

Provided there are only two alternative food sources, and that their $\delta^{13}C$ signatures are known, and differ significantly from each other, $\delta^{13}C$ analysis can be used to ascertain the dominant food source, and to estimate its relative contribution. By comparing $\delta^{13}C$ values of different organisms, it is also possible to track mangrove carbon through successive stages of food chains. Figure 6.4 shows the distribution of $\delta^{13}C$ values of samples of mangrove and algal material, and in the tissues of various other mangrove animals. Only in gastropod molluscs and a few crabs (mostly sesarmids) do the $\delta^{13}C$ values overlap with those of mangroves, confirming the importance of mangrove material in the diet of these species (Rodelli *et al.* 1984).

Are crabs selective feeders?

In most mangrove areas there is a standing crop of leaf litter available to a foraging crab. This is made up of fallen leaves varying in size, physical toughness, nutritional value, and palatability, depending on the species of tree from which they came, and the time for which they have been decomposing after their fall. Do crabs feed selectively, or indiscriminately, from this varied menu?

Preferences have been established in the crab *Neosarmatium smithi*, a large sesarmid abundant in the mangroves of Queensland, Australia. Freshly picked and senescent leaves from *Ceriops tagal* trees were allowed to age on the mud surface for different periods of time, before being presented to crabs in different combinations. After 24 h, the leaf fragments that remained were sorted into their original categories and the amount of each category eaten was worked out by subtraction from the amount present at the start of the experiment. When the choice was between fresh, senescent, and decayed leaves, crabs clearly preferred decayed (Table 6.2). In further experiments, crabs were fed each leaf type separately—newly picked, and decayed for various periods of up to 10 weeks—and the results expressed as the amount eaten relative to the total for all diets. Figure 6.5 shows the increasing relative consumption rate as the leaves progressively decay. Why are decayed leaves preferred to fresh or senescent ones? As decay proceeded, the leaf content of flavolans decreased. Flavolans, or condensed tannins, are

Table 6.2 Consumption rate of *Neosarmatium smithi* when offered a choice of decayed, fresh, or senescent leaves of *Ceriops tagal*. From Giddins et al. (1986), with permission of Inter-Research Science Center.

Leaf type	Consumption rate (g of leaf wet weight/g of crab per day; ±S.D.)
Fresh	0.004±0.006
Senescent	0.011±0.014
Decayed	0.062±0.055

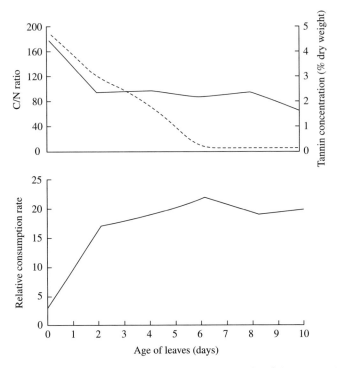

Fig. 6.5 Upper graph: changes in tannin concentration (percentage of leaf dry weight; dashed line) and C/N ratio (solid line) of leaves of *Ceriops* at different stages of decomposition. Lower graph: relative consumption rate by *Neosarmatium smithi* of leaves of *Ceriops* at different stages of decomposition. Data, with permission, from Giddins, R.L. Lucas, S.J., Neilson, M.J., and Richards, G.N. 1986. Feeding ecology of the mangrove crab *Neosarmatium smithi* (Crustacea: Decapoda: Sesarmidae). *Marine Ecology Progress Series* **33**: 147–155, with permission of Inter-Research Science Center.

present in relatively high concentrations in mangrove leaves, for the most part covalently linked to glycan carbohydrates as flavoglycans. They make leaves unpalatable or indigestible. The leaching out of flavolans correlates with the increasing relative consumption rates shown in Figure 6.5. This could, of course, be coincidental, but when flavolans are artificially added to otherwise palatable decayed *Ceriops* leaves, the amount eaten was markedly reduced. Leaf flavolans deter feeding (Neilson *et al.* 1986).

Change and decay also alter the nutritional value of mangrove leaves. A useful measure is the ratio of carbon to nitrogen (C/N ratio). The higher this ratio, the less valuable the food. A C/N ratio below 17:1 is regarded as being necessary for a nutritious food. Figure 6.5 shows that the C/N ratio tends to decrease with time: the older the leaf, the better. Even at their best, however, rotting leaves appear to be an inadequate diet, and it is hard to see why crabs eat them. Even sediment detritus would appear to be better value than leaves (Skov and Hartnoll 2002). Probably leaves are an adequate source of carbon, and crabs depend critically on occasional ingestion of animal tissue for their nitrogen (Thongtham and Kristensen 2005).

Where more than one species of mangrove tree occurs, crabs will have the choice of fallen leaves of different species, as well as leaves at different stages of decomposition. In some cases, crabs have been shown to prefer leaves of one species over those of another. Senescent leaves from three mangrove species were offered to two sesarmids common in the Mai Po marshes of Hong Kong, *Perisesarma bidens* and *Parasesarma maipoensis*. The order of preference was *Avicennia marina*, followed by *Kandelia candel* and then *Aegiceras corniculatum*. This correlates with the order of increasing C/N ratios: nutritionally, *Avicennia* is better value, followed by *Kandelia*, then *Aegiceras*. *Avicennia* also has the lowest content of aversive tannins (Table 6.3). An Australian sesarmid, *Parasesarma erythrodactyla*, has similarly been shown to prefer *Avicennia* to other species of mangrove (Camilleri 1989). Growth and survival of sesarmids appears also to be better on a diet of *Avicennia* than *Kandelia*, and to be better on decayed leaves (with lower C/N ratios) than on fresh (Kwok and Lee 1995).

In contrast, studies in East Africa showed that *Avicennia marina* was the *least* preferred of five species of mangrove leaf, the sequence being *A. marina*<*C. tagal*<*Rhizophora mucronata*<*Sonneratia alba*<*Bruguiera gymnorrhiza*. Two species of crab were involved: *Neosarmatium meinerti* and the gecarcinid land crab *Cardisoma carnifex*. Both showed the same

Table 6.3 Relative preference of two sesarmid species, *Perisesarma bidens* and *Parasesarma maipoensis*, for leaves of three mangrove species, compared with C/N ratios and soluble tannin content of the leaves. Preferences are the mean ranking for both species and both sexes combined, and percentage soluble tannins are mean values±S.D. From Lee (1993).

Mangrove species	Preference rank	C/N ratio	Percentage of soluble tannins
Avicennia marina	1.30	27.4	0.86±0.03
Kandelia candel	1.89	49.1	2.35±0.29
Aegiceras corniculatum	2.80	69.1	1.95±0.19

order of preference, although only with *C. carnifex* were the results statistically significant. The experiment was carried out in high intertidal forests: it may be that the apparently perverse order of preference was due to the need to obtain scarce water from more succulent leaves, the water content of *Avicennia* leaves being relatively low (Micheli *et al.* 1991).

Although not all investigators have shown selective feeding, these experiments suggest that crabs may choose one species of fallen leaf over another, and in general prefer decaying leaves to fresh ones. In either case, the preference could be on the basis of tannin concentration or nutritional value.

Crabs do not always eat leaves on the surface: they may instead take them down burrows. After a week or two, tannin levels and C/N ratios fall to more acceptable levels. This behaviour makes sense from the crab's point of view, provided it does not abandon its burrow and contents in the mean time. In some areas, the amount of leaf material just below the mud surface suggests that crabs are not very efficient in dealing with buried leaves. Sometimes leaves are eaten almost immediately after their disappearance below the surface, and before any appreciable change has taken place in tannin levels; one estimate was that 78% of leaf material was eaten within 6 h of burial (Robertson 1988). Crabs may therefore take leaves underground for other reasons than to store them until they decay and become more palatable.

What other advantages might there be in taking food into burrows? Leaf-burying, or caching, behaviour will sequester food resources, making them unavailable to competing crabs and other herbivores. Alternatively, burying leaves will prevent them from being washed out by the tide and extend the time when they can be fed on. If so, caching would be more common where tidal outwash was more likely. Finally, burying leaves will minimize the time a foraging crab spends on the surface, exposed to predators, high temperature, or other uncongenial circumstances. Leaf burial may therefore be more common where there is greater predator pressure, or more intense competition between crabs for a limited supply of leaves, or under different tidal-inundation régimes (Giddins *et al.* 1986; Camilleri 1989). A crab's life is a complex balance of opportunities and risks, costs and benefits, and how it allocates its time can be crucial to survival (p. 122).

Seedlings

Mangrove propagules or seedlings are a major food source for sesarmid crabs. In Australia and Malaysia, the majority of propagules are destroyed within days of their release from the tree (and the released propagules are themselves the few survivors of pre-release attack by insects and other causes of death; see p. 34). Crabs show preferences for different species. When propagules of five species of mangrove in tidal forests in north Queensland were tethered and their survival rates monitored over 18 days,

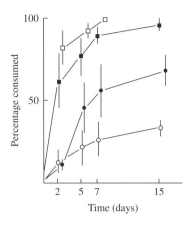

Fig. 6.6   Cumulative predation of *Aegiceras* seedlings. ○, ●, Low intertidal; □, ■, high intertidal; ○, □, open spaces; ●, ■, forest understorey. From Osborne, K. and Smith, T.J. 1990. Differential predation on mangrove propagules in open and closed canopy forest habitats. *Vegetatio* **89**: 1–6, Figure 3 (part), with kind permission from Springer Science and Business Media.

there were clear differences in the rate of removal by crabs. Virtually every *Avicenia marina* seedling was destroyed by crab attack within the experimental period. Preference was closely related to nutritive quality, with *Avicennia* propagules having the highest content of simple sugars and lowest levels of tannins, fibre, and protein (Smith 1987c).

The extent of seedling predation depends on other factors besides nutritional quality. Predation is generally greater at higher levels on the shore, reflecting shorter submergence times and consequently more time available for foraging (Figure 6.6). The greater the crab predation on a species, the less that species is likely to be a dominant constituent of the forest. In Queensland mangrove forests, *Avicennia* is present at the upper and lower shore, but absent from mid shore. *Avicennia* propagules do strand here, and there is nothing physically to prevent the species from establishing itself: if crabs are artificially excluded then it does so. The 18-day predation rate of tethered *Avicennia* seedlings in the midshore zone was 100%, suggesting that the absence of *Avicennia* here is due to the presence of crabs (Smith 1987c). Predation is also more intense in thick mangrove forest than where there are gaps in the canopy. The bigger the gap, the lower the rate of seedling predation (Figure 6.7). This probably also relates to crab foraging behaviour. Either the extra travelling time in the open makes foraging less cost-effective, or it increases a crab's vulnerability to predators.

Although many other factors are no doubt involved, crabs appear to play a major role in litter processing, and in determining the local distribution of mangrove tree species and community structure of mangrove habitats.

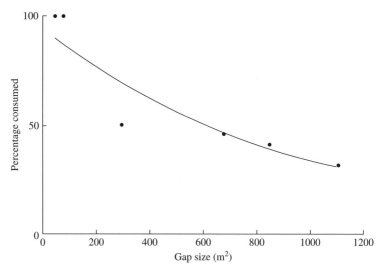

Fig. 6.7 Seedling predation by crabs (as percentage eaten) in relation to forest gap size. From Osborne, K. and Smith, T.J. 1990. Differential predation on mangrove propagules in open and closed canopy forest habitats. *Vegetatio* **89**: 1–6, Figure 2, with kind permission from Springer Science and Business Media.

How important are herbivorous crabs?

Overall, sesarmids play a very important part in the ecology of mangroves, at least in Australia and Malaysia. Selective destruction of propagules determines the distribution pattern of tree species within the mangal, while processing of leaf litter expedites breakdown and makes this crucial energy source more readily available to other organisms. One estimate is that litter processing by crabs increases litter turnover to more than 75 times the rate that would obtain with microbial decay alone (Robertson and Daniel 1989). Is this typical of all mangrove ecosystems?

In the mangroves of south-west Florida, the crab fauna is dominated, not by herbivorous sesarmids, but by deposit-feeding ocypodid crabs, and predatory crabs belonging to the family Xanthidae. Experiments with tethered leaves showed no removal by herbivorous crabs. There is some evidence that leaf eating by crabs does take place in waterways after the leaves have been flushed out of the forest. Elsewhere in the Caribbean, leaf processing does take place: turnover decreases with height above low tide, the opposite of the situation in Australia. There appear to be important and interesting differences between the Caribbean and the Indo-Pacific in the ecological significance of mangrove crabs (Robertson and Daniel 1989; McIvor and Smith 1996).

Seedling predation by crabs also varies geographically. In Florida, overall rates of removal by crabs are, as would be expected, much lower: 6% of

Avicennia germinans seedlings were taken in Florida, compared with 95–100% of *A. marina* in Australia and Malaysia. For the less desirable *Rhizophora*, the figures were similar in Australia and Malaysia: none were taken in Florida. Gastropod molluscs are more important herbivores than crabs in Florida mangroves, accounting for 73% of seedlings, as opposed to 0–5% in Australia and Malaysia.

The relationship between tree dominance and crab predation of seedlings has also been investigated in Caribbean mangrove communities. In Belize, predation of *Avicennia*, as in Australia, was higher where the tree species was least dominant. The major seedling predators were the grapsid *Goniopsis* and the hairy land crab *Ucides* (Ocypodidae). In contrast, however, consumption of the seedlings of *R. mangle* and of a third mangrove species, *Laguncularia racemosa*, was highest where the adult trees were most, not least, dominant (McKee 1995). The importance of crabs in mangrove community structure and function therefore differs from one part of the world to another.

Tree-climbing crabs

Crabs are not restricted to scavenging on the mud surface. Several species actually climb trees. In Malaysian mangroves, for example, the large sesarmid *Episesarma versicolor* climbs a metre or two up tree trunks, particularly at night. These crabs appear to be scraping diatoms or other microphytes off the tree bark, but may also be avoiding aquatic predators such as fish or otters. They are certainly very vigilant where terrestrial predators are concerned, and drop off into the water if a human approaches closer than 2 or 3 m. *Episesarma* is not truly arboreal. In the Caribbean, East Africa, and West Africa are three sesarmid species which spend virtually all of their adult life in trees: *Aratus pisonii*, *Armases elegans*, and *Parasesarma leptosoma*, respectively. Despite their similarities, these species are not particularly closely related, and represent a striking instance of convergent evolution (Fratini *et al.* 2005). The best known is the Caribbean *A. pisonii* (Figure 6.2b). *Aratus* has short dactyls, used more or less like the pitons of a mountaineer. It is a very agile species and can achieve speeds of about 1 m/s. At night it moves up to the tops of trees, perhaps 10 m above ground, and feeds on buds and young leaves.

The East African species, *P. leptosoma*, virtually identical to *Aratus* in appearance, has a similar way of life. As an adult, this species never enters the water, nor does it even venture onto the mud surface. Much of the day is spent foraging on the aerial roots of the host tree (usually *R. mucronata*, *B. gymnorrhiza*, or *C. tagal*) and, presumably, replenishing water. Twice a day crabs migrate in synchrony to the treetops. At about 06.00 h the 200–500 crabs occupying each tree move upwards, returning about 4 h later. A second migration follows at around 17.00 h, with the return at dusk. While in the treetops the crabs feed by scraping tissue off the lower surface of mature leaves. They also spend much of their time on leaf buds at a particular stage of development. Besides food, the buds may accumulate valuable water. Individual crabs faithfully return to the same feeding sites by the same route, and return to the same root crevices on their return (Vannini and Ruwa 1994; Cannicci *et al.* 1996a, 1996b).

The remarkable synchrony of the vertical mass migrations may reduce predation. Fidelity to feeding and refuge sites may also give the crabs an advantage in avoiding predators through familiarity with the terrain. Alternatively, a fixed feeding area may improve the chances of finding patchily distributed leaf buds. It is not clear, however, what particular combination of circumstances was conducive to the evolution of such complex behaviour in this particular species. Nor is it known how *P. leptosoma* copes with the problems of high tannin and C/N ratios which seem to have shaped the very different behaviour of other sesarmids.

Fiddler crabs

The most colourful mangrove invertebrates are crabs of the family Ocypodidae; among these, the most striking are undoubtedly the fiddler crabs (*Uca* spp.), which construct burrows at all levels of the shore. Fiddlers have bodies that are broad in relation to their length, and almost circular in midline section, so that they fit neatly, sideways, into a burrow. Males have one greatly enlarged claw, used in social displays and in jousting with rival males (Figure 6.8). Particularly during displays, the carapace and claw are often very brightly coloured in crimson, orange, or intense blue, depending

Fig. 6.8   A male fiddler crab (*Uca* sp.). Photo: D. Barnes.

on species. Females are much less flamboyant, and have two claws of normal size.

Like many other ocypodids, fiddler crabs are deposit feeders, ingesting organic matter from the exposed mud at low tide. Sediment is carried to the mouthparts by the minor chela in the male, and by both chelae in the female. The tips of the chelae are generally expanded, almost spoon-shaped, and often fringed with setae (Figure 6.9). Sediment is sorted within the buccal cavity. The outermost mouthparts, the third maxillipeds, play little active part other than helping to retain sediment and water during the sorting process. The inner surface of the second maxillipeds carries quite large numbers of long setae, some with spoon tips and others feathery. Facing these on the outer surface of the first maxillipeds is a brush of stiff setae. The sediment is rolled between two maxillipeds. The spooned setae of the second maxillipeds hold sand grains against the brush-like setae of the first maxillipeds, and diatoms and bacteria adhering to the grains are brushed off and moved towards the mouth itself. While this is going on, water is pumped out of the gill chamber into the buccal chamber. This helps the sorting process which takes place essentially in suspension. Sand grains, being denser, fall back towards the third maxillipeds. Together with unwanted organic matter, and some mucus from the crab's saliva, they are formed into so-called pseudofaecal pellets and discarded. Much of the water is recirculated into the gill chambers, mainly through an aperture near the base of the outer, third maxilliped. Clogging of the gill surface with sediment particles is prevented by an array of filtering setae protecting this aperture.

Sediments vary in particle size and organic content, and different species of *Uca* have specialized in different sediments. Species which prefer coarser sandy sediments have chelae with a wider gape, to accommodate larger particles. Those which specialize in finer sediments tend to have a narrower gape, and to have the tips of the fingers of the claw fringed with setae to retain mud as it is carried towards the mouth (Figure 6.9). The mouthparts, too, are adapted to particular sediments. Species feeding predominantly on coarse sandy sediments have more of the long spoon-tipped setae on the inside of the second maxillipeds, and the tips of the setae are more spoon-shaped, and the setal brush on the outside of the first maxillipeds is denser. Where the preferred sediment contains more fine organic particles, extra rows of setae are present at the base of the third maxillipeds to protect the aperture into the gill chamber and prevent the gills from becoming clogged (Miller 1961; Ono 1965; Macnae 1968).

Different food preferences are reflected also in the structure of the masticating structures in the foregut. Crabs have a single median tooth here, the surface of which carries a pattern of hard ridges. Two lateral teeth, with rows of short stiff setae, are brought to bear on this surface. In species

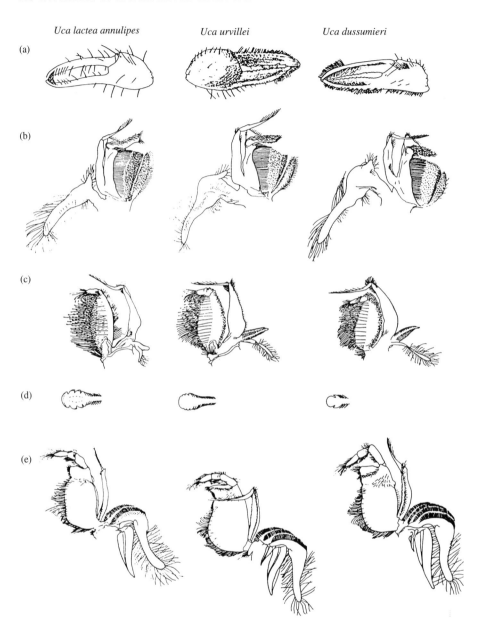

Fig. 6.9 Adaptations of fiddler crabs (*Uca* spp.) to different soil particle sizes. The three columns represent *Uca lactea*, *Uca urvillei*, and *Uca dussumieri*, respectively. *Uca lactea* lives on coarser sediments than the other two species. (a) Feeding chela (minor chela of male, either chela of female). (b) Outer view of first maxilliped. (c) Inner view of second maxilliped. (d) Spoon-shaped setae from second maxilliped. (e) Outer view of third maxilliped. For description of how the mouthparts operate, see text. Reproduced with permission from Macnae, W. 1968. A general account of the fauna and flora of mangrove swamps and forests in the Indo-West Pacific Region. *Advances in Marine Biology* **6**: 73–270, with permission from Elsevier; and Warner (1977).

such as *Uca lactea* the median tooth has parallel ridges forming an extremely abrasive surface suitable for dealing with the coarse sandy sediments which this species prefers. In species that process finer, softer particles, the hard ridges on the median tooth are reduced and partly replaced by stout hooked setae, whereas the setae of the lateral teeth are longer and more flexible (Icely and Jones 1978).

The sophisticated sorting mechanism does more than just separate inorganic from organic particles. In two species of Australian fiddler crabs, *Uca vocans* and *Uca polita*, bacteria and protozoa were found to be the most abundant cells in the gut, which contained relatively few algal cells. In the rejected material of the pseudofaecal pellets, the position was reversed, showing that the buccal sorting mechanism can select one type of cell and reject another.

Selection continues within the gut. When the same two species of *Uca* were fed with radioactively labelled bacteria, radioactivity of the faeces was a tiny fraction of that of the gut contents, showing that bacteria were efficiently digested. Assimilation efficiency was estimated at greater than 98% in both species. The assimilation efficiency for algal cells was 41 and 31% in *U. vocans* and *U. polita*, respectively. In general, deposit feeders such as *Uca* seem to depend much more on bacteria, diatoms, and the smaller members of the meiofauna such as ciliate protozoa and nematodes than they do on larger algal cells and detrital material (Dye and Lasiak 1986, 1987). Stable isotope analysis confirms that they do not depend on mangrove detritus, or do so only indirectly.

Because only a small proportion of the total sediment mass is selected as food, fiddler crabs must process large amounts of sediment. During periods of active feeding, the chelae are almost continuously active. The enlarged display chela of the males is useless for feeding, so males must work harder with the single minor chela, and have lower overall intake rates. Poorer soils require a higher rate of intake: on a muddy Malaysian mangrove shore, rich in organic matter, male *Uca dussumieri* fed at the rate of 25–60 scoops/min while *U. lactea annulipes*, on a more sandy shore with lower organic content, scooped at a rate of 60–140 scoops/min (Macintosh 1988). Smaller and more local differences in soil organic content, detected by probing with the minor chela, may influence the behaviour of individual crabs. A calculating crab should abandon a particular patch of mud and move to another whenever its current rate of energy intake falls below a certain threshold of cost-effectiveness. Measurement of feeding costs, in terms of oxygen-consumption rates, and benefits, in energy value of the food acquired, makes it possible to predict this leaving threshold. The predicted threshold value will be different for males and females, as they have different intake rates and foraging costs, and at different concentrations of food in the substrate. At least one species of (salt-marsh) fiddler crab investigated, *Uca pugnax*, behaves exactly as predicted. Fiddler crabs are therefore capable of optimizing their foraging strategies to match a heterogeneous environment (Weissburg 1993).

Most feeding takes place within a few hours of the tide receding, while the mud surface is still wet and easily handled. Later, when the surface has dried, is the time for more social activity. The allocation of appropriate time to different activities, as well as selection of where to forage, therefore promotes efficiency.

Fiddler crabs are often very abundant: densities of 70 crabs/m^2 are not unusual in south-east Asian mangroves (Macintosh 1982). Each crab may produce around 300 pseudofaecal pellets a day, with a dry weight of around 36 mg each. Before the returning tide disperses them, pseudofaecal pellets sometimes almost entirely cover the soil surface. In addition, a fiddler crab may produce some 25 faecal pellets of 8 mg each. Compared with the unaltered substrate, the nitrogen and carbon content of these is considerably enriched. Allowing for some of the material being processed more than once in the course of a day, the turnover rate of soil may be of the order of 500 g/m^2 per day. Fiddler crabs have a major role in retexturing the soil and altering its chemical composition, and in doing so significantly modify the soil environment (Ono 1965; Macintosh 1982). Burrowing also has a major physical impact on the environment, discussed below (p. 125).

Fiddler crabs are an important component of the mangrove ecosystem in another way: as food for numerous predators. These include mammals (raccoons, monkeys), birds (kingfishers, herons), reptiles (snakes), fish (mudskippers), and other crabs. The annual production of fiddler crabs on a Malaysian mangrove shore has been estimated at 3.1–16.0 g/m^2, equivalent to 4.6–28.9 kcal/m^2, most of which is accounted for by predation (Macintosh 1982).

Reproductive adaptations

As with almost all species of crab, fiddler crabs have planktonic larvae. These too are a major food resource for predators, in this case mainly fish and other inhabitants of mangrove creeks and offshore waters. Crab larvae are minute, only 1 mm or so in length, but are copiously produced. One female *Uca rosea* (a common small south-east Asian species) has a brood size of 10 000: with around 4.5 batches a year, this gives an estimated annual production of 45 000 larvae. Taking into account the recorded population densities of fiddler crabs, and allowing for the proportion of mature females in the population, annual production may be of the order of half a million larvae per square metre of mangrove. A mangrove creek in Costa Rica, 325 ha in area, was conservatively estimated to export 7.5×10^{10} *Uca* larvae annually (Macintosh 1982; Dittel *et al.* 1991).

Females may synchronize the release of their larvae as a way of swamping predators and minimizing losses, but only an infinitesimal fraction of the

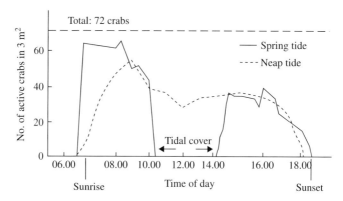

Fig. 6.10 Pattern of activity of *Uca lactea* on a mangrove shore in Thailand, in relation to time of day and state of the tide. Reproduced with permission from Macintosh, D.J. 1988. The ecology and physiology of decapods of mangrove swamps. *Symposia of the Zoological Society of London* **59**: 315–341.

larvae released survive to settle on the shore and metamorphose into juvenile crabs.

Fiddler patterns in time and space

Mangrove fiddler crabs are active only during daytime, at low tide (Figure 6.10). At other times, they retreat into burrows. Other considerations aside, therefore, it would be advantageous to live high on the shore since this allows more time for foraging and other activities. On the other hand, the particle-sorting method of feeding requires water, and although water is recirculated between buccal cavity and gill chamber, losses occur and must be replaced. Standing water is more readily available lower down the shore, so a crab foraging in the lower shore will not need to return to its burrow so frequently to replenish its water supply, and will be able to forage further afield.

The different species of *Uca* are adapted to feeding on different soil textures, and soil texture varies at different levels of the shore. Feeding methods therefore interact with other factors in determining the pattern of vertical zonation shown by fiddler crabs on a mangrove shore, an example of which is shown in Figure 6.11. Other physiological problems which affect distribution are shared with sesarmids (and, indeed, with the majority of mangrove-living invertebrates), and are discussed next.

The physiology of living in mud

The problems facing mangrove crabs are much the same as those solved by mangrove trees themselves: coping with low oxygen concentrations in water-logged mud, conservation of water, and withstanding high temperatures and varying salinity.

COPING WITH LOW OXYGEN LEVELS Most mangrove crabs are active at low tide. While the tide is in and the mud surface submerged, they retreat

120 THE BIOLOGY OF MANGROVES AND SEAGRASSES

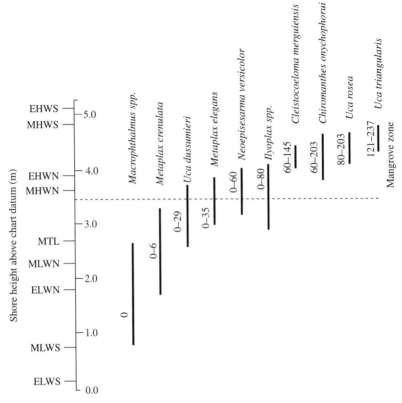

Fig. 6.11 Vertical zonation of crabs on a Malaysian mangrove shore. Numbers shown for each vertical range are days per year without being covered by the tide. *Macrophthalmus*, *Uca*, and *Ilyoplax* belong to the family Ocypodidae, the remainder are Grapsidae or Sesarmidae. EHWS, MHWS, MLWS, ELWS, extreme high water, mean high water, mean low water, and extreme low water spring tides; EHWN, MHWN, MLWN, ELWN, the same, neap tides; MTL, mean tide level. Reproduced with permission from Macintosh, D.J. 1988. The ecology and physiology of decapods of mangrove swamps. *Symposia of the Zoological Society of London* **59**: 315–341.

into burrows and remain inactive. Oxygen concentration within the burrow is low. In laboratory conditions, *Uca* can survive anoxic conditions for up to 40 h. During this time the crab accumulates lactate within its body fluids. When it returns to normal conditions, oxygen consumption is raised for some time to eliminate the accumulated lactate. Fiddlers survive anoxia by building up an oxygen debt. Sesarmid crabs probably have similar tolerance of low oxygen levels. Sesarmids caught overnight in flooded pitfall traps, in a hot Malaysian mangrove forest, are very often found dead. This may indicate limited tolerance of anoxia; or, perhaps, of high temperatures.

Most of the activity of fiddler crabs, and sesarmids, takes place in air. The oxygen concentration of air, at 20%, is considerably higher than in water.

Offsetting this is the fact that gills in general are not well adapted to air-breathing. Finely divided respiratory tissues tend to collapse under surface tension, reducing their effective surface area for gas exchange. Accordingly, the gill area is very much reduced compared with that of fully aquatic crab species, and the walls of the branchial chamber are vascularized for gas exchange, in an approximation to a vertebrate lung (Gray 1957; Takeda *et al.* 1996).

Sesarmid crabs have a supplementary mechanism. The ventral carapace plates on either side of the mouthparts and below the eyes carry a very distinctive array of short bristles. Water is pumped out through exhalant openings at the side of the buccal cavity, is spread as a thin film by the network of bristles, and is then taken back into the gill chambers again at an inhalant opening near the base of the legs (Figure 6.12). Gas exchange with the atmosphere takes place during this recycling process. Some water passes over the crab's dorsal surface, where it is again spread out into a thin film by lines of short bristles. Here, too, gas exchange occurs (Felgenhauer and Abele 1983).

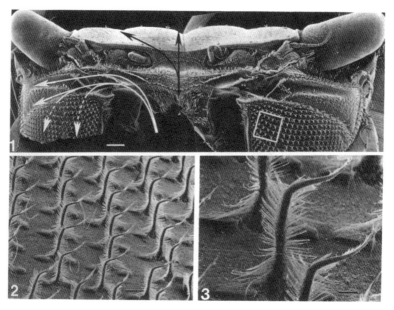

Fig. 6.12 The reticulated network of short setae, and the route of water recirculation, on the under surface of a sesarmid crab, *Sesarma reticulatum*. (1) Frontal view of the crab. White and black arrows indicate the path of respiratory water across the the carapace (scale bar, 1 mm). (2) Close-up views of the seta-covered plate to show arrangement of setae (scale bar, 100 μm). (3) Further magnification to show presence along shaft of seta of fine setioles for water retention (scale bar, 50 μm). Reproduced with permission from Felgenhauer, B.E. and Abele, L.G. 1983. Branchial water movement in the grapsid crab *Sesarma reticulatum* Say. *Journal of Crustacean Biology* **3**: 187–195. ©1983 by The Crustacean Society.

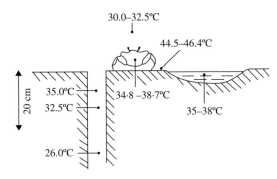

Fig. 6.13 Temperature at various points occupied by a fiddler crab (Uca). Reproduced with permission from Edney, E.B. 1961. The water and heat relationships of fiddler crabs (Uca spp.). *Transactions of the Royal Society of South Africa* **36**: 71–91.

HIGH TEMPERATURE The temperature at the mud surface in a tropical mangrove swamp can rise to 45°C, close to the lethal limit for most animals. Fiddler crabs can keep their body a critical few degrees below the prevailing air temperature by evaporation. The external recirculation of water by sesarmids cools the crab by evaporation, as well as facilitating gas exchange. For the most part, however, mangrove crabs probably maintain a reasonable body temperature by movement in and out of burrows. Figure 6.13 shows the temperatures measured at various locations in which a fiddler crab (*U. lactea*) spends its time. As the temperature rises, a crab makes more frequent trips back to its burrow and spends proportionately less time on feeding and other activities (Edney 1961; Eshky *et al.* 1995).

WATER ECONOMY It may seem strange to suggest that animals living in a mangrove swamp might have a problem with water shortage. In some mangrove areas, the tide goes out a considerable distance horizontally, and pools of standing water are not necessarily abundant. Frequent returns to a burrow to replenish water take time that could be devoted to feeding or social behaviour.

Some species are relatively impermeable to water. Water passes through the carapace of *Uca* at rates of about 3 mg/cm^2 per h, and *Sesarma* at 6 mg/cm^2 per h, compared with a range of 14–21 mg/cm^2 per h for a variety of non-mangrove species (Vernberg and Vernberg 1972, quoted in Hutchings and Saenger 1987). This reduction in water loss through the carapace is at the expense of evaporative cooling. Some species, including *Sesarma*, have an additional means of gaining water from the soil by what are, in effect, roots. Tufts of hydrophilic setae at the bases of the legs are brought into contact with the moist surface of the mud and can actually draw water into the crab's body (Burggren and McMahon 1988).

OSMOTIC PROBLEMS The final problem to consider is that of varying salt concentrations, specifically, the osmotic imbalances that might result.

Crabs, of course, are marine in origin: their starting point is therefore the salinity of full sea water, in contrast to mangrove trees which are of terrestrial origin and come from ancestors accustomed to fresh water.

Only a few osmoregulatory strategies are possible. An animal could be an *osmoconformer* and allow its body fluids simply to equilibrate in osmotic concentration with the external medium. This saves energy that would otherwise be devoted to osmoregulation by actively transferring water in or out of the body. Perfect osmoconformity over the full range of salinities is impossible: at very low external concentrations, tissue fluids and cell contents would have to approach the composition of pure water, while extremely high ionic concentrations interfere with protein conformation and physiological functions. Perfect *osmoregulation* would be expensive in a small animal (with a high surface area/volume ratio) without a completely impermeable integument. This extreme strategy is therefore not available to crabs. Between these extremes, various compromises are possible. An animal could maintain an osmotic concentration greater than that of the environment (hyperosmoregulation) at low osmotic concentrations, but conform at higher ones, or hyperosmoregulate at lower concentrations and hyporegulate at higher ones. Among mangrove crabs, the commonest strategies involve hyperosmoregulation at low external concentrations, and a varying degree of hyporegulation at higher concentrations (Figure 6.14).

How stressful is a crab's life?

If a crab is turned on its back and released, it soon struggles back to the correct position and scuttles off. The healthier the crab, the quicker it is to do this. If the crab has been suffering some form of physiological stress, its physiological responses are slower and co-ordination poorer, and it is slower to right itself. Measuring the righting response time (RRT) can therefore give a quantitative assessment of the degree of physiological stress to which a crab has been subjected.

This method has been used to judge how stressful conditions in mangrove habitats are to crabs, by measuring the righting response time after crabs had been restrained for periods of 3 h on the mangrove mud, among mangrove roots, or in the mangrove canopy. Controls were freshly caught crabs. By this criterion, the most stressful situation was in the mangrove canopy, followed by the mangrove roots, then the open mud. Least stressed were the freshly caught controls (Wilson 1989). This is not surprising, although it would be interesting to compare stress levels in the canopy for a habitually tree-climbing species (p. 113). It does raise the question of why crabs ever leave the least stressful environment of open mud. Crabs are more vulnerable to predation on the open mud than among mangrove roots. Physiological stress and predation risk are being traded off against each other.

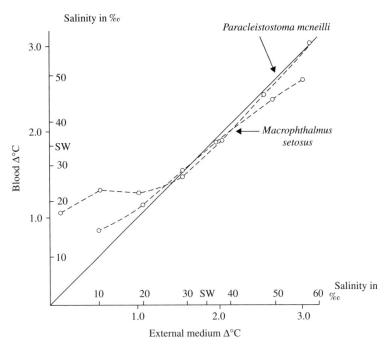

Fig. 6.14 Osmotic concentration of the blood of some ocypodid crabs in varying external osmotic concentrations. Osmotic concentration is given as freezing point depression ($\Delta°C$); the diagonal line represents complete osmotic conformity with the external medium. SW, sea water, approximately 35‰. Reproduced with permission from Barnes, R.S.K. 1967. The osmotic behaviour of a number of Grapsoid crabs with respect to their differential penetration of an estuarine system. *Journal of Experimental Biology* **47**: 535–551. ©Cambridge University Press.

A crab's life is one of compromise and trade-offs: of one physiological stress against another, of predation risk against physiological stresses, of the need to feed efficiently, and social and reproductive requirements against all of these. Particular patterns of distribution and activity in time and in space are the result of these conflicting requirements.

6.3.1.2 Other mangrove crustacea

Grapsoid and ocypodid crabs, particularly sesarmids and *Uca*, dominate the mud surface at low tide. As we have seen, these are predominantly herbivores and deposit feeders, respectively, although some members of the families are relatively omnivorous. Other crabs found in mangroves may be important predators. The most conspicuous is the mud crab *Scylla*, of the family of swimming crabs (Portunidae). *Scylla* reaches a carapace width of up to 20 cm, making it the largest invertebrate predator found in mangroves. Equally formidable predators are the mantis shrimps (Stomatopoda) which live in burrows in the mud, and lacerate prey by rapidly shooting out their spiked raptorial appendages: at least one species, *Squilla choprai*, is found in mangroves in Malaysia (Macnae 1968).

More general mangrove scavengers include hermit crabs, particularly *Clibanarius* species, which forage on the mud surface at high tide. Some hermits may also climb into the mangrove trees. Shrimps and prawns (the terms are used interchangeably) may also be abundant in mangroves and mangrove creeks. Penaeid shrimps, which in at least some parts of the world depend heavily on mangroves for feeding and breeding, are an important commercial crop. The relationship of mangroves to shrimp fisheries and aquaculture is discussed below (p. 178).

The shrimp *Merguia* apparently lives only in mangroves, and has the distinction of being the only semi-terrestrial shrimp known: it actually climbs trees. Only two species are known. One occurs in the Indo-West Pacific region, from Kenya to Indonesia; the other has been found in Panama, Brazil, and Nigeria (Bruce 1993). The Indo-West Pacific and Atlantic regions differ in the composition of their mangrove floras (see p. 185). The divergence of the two species of mangrove-associated shrimps presumably paralleled the divergence of the mangroves themselves. This contrasts with the convergent evolution of the tree-climbing crabs.

Numerous Crustacea live under the mud surface. Pistol shrimps (*Alpheus*) are more often heard than seen: they live in burrows in the mud, and have one greatly enlarged claw, used to make the loud snapping sound which gives them their common name. Also seen only rarely at the surface is the mud lobster (*Thalassina*), a subterranean deposit feeder which ejects waste mud into sizeable mounds (Macnae 1968).

6.3.1.3 *Crustaceans as ecosystem engineers*

All species have an impact on their environment: at the very least, they exchange materials in the form of food, waste materials, and respiratory gases. Some species have an effect beyond these straightforward transactions, and alter the nature of the environment in ways that affect species other than their direct competitors, predators, or prey. Such species are often termed ecosystem engineers. Ecosystem engineers are species that 'directly or indirectly modulate the availability of resources to other species, by causing physical state changes in biotic or abiotic materials. In so doing they modify, maintain and create habitats' (Jones *et al.* 1994).

The trees themselves are the greatest of the ecosystem engineers in mangroves. Self-evidently, they create the mangrove habitat, and by their physical nature and organic production greatly modify the muddy soil in which they grow. Mangrove crustaceans, and particularly crabs, are also significant ecosystem engineers. The topography of mangrove swamps in south-east Asia is often visibly modified by mud lobsters (*Thalassina*). In the processing of burrowing, and extracting organic matter from the mud,

Thalassina throw up waste material from beneath the surface. This accumulates as mounds which are often well over 1 m in height, and may even reach 2 m. *Thalassina* mounds create large patches of relatively dry mud which provide suitable habitats for a variety of other species, including the mangrove fern *Acrostichum*, fiddler crabs, and the large sesarmid *Episesarma*, as well as other burrowing crustaceans and molluscs. Between the mounds the mud is more frequently inundated by the tides, and more waterlogged, than it would otherwise be. Burrowing crabs also contour their environment, although not so dramatically (Warren and Underwood 1986).

Crab activities also have less dramatic effects. Much of the microbial activity of mangrove mud occurs within a short distance of the surface, to a depth limited by diffusional gas exchange with the atmosphere. This activity will be enhanced by fiddler crabs working over the surface as they extract food, continually exposing fresh mud to the surface and sorting soil particles by size and composition. Burrowing has the further effect of increasing the surface area of mud exposed more or less directly to the atmosphere. A square metre of mangrove mud may contain 40–50 crabs. If each crab constructs a burrow 10 cm deep and 1 cm in diameter, the result is to increase the surface area of the mud by around 12% (quite apart from the effects of redistributing the material excavated from the burrows during construction). This simple calculation may underestimate the importance of crab burrows: an analysis of the effects of fiddler crabs in a saltmarsh indicated that burrowing increased the surface area by 59% (Katz 1980).

Although some crab burrows are relatively simple in structure, with exclusive occupancy by a single crab, others are not. Mangrove mud is often honeycombed with a network of interconnecting passages, more or less promiscuously occupied by a shifting population of crabs. The labyrinthine state is rarely obvious from the surface of the mud, but is revealed at low tide by the amount of water that flows out of crab holes in the banks of mangrove creeks, and by the observation that the rising tide often wells out of crab burrows long before it starts advancing across the surface of the mud. Dye injection confirms the extent of water movement under the mud surface, through crab burrows (Ridd 1996).

Are effects such as these of any importance? Fiddler crabs are known to influence the productivity of saltmarsh vegetation: reduction in crab density in small experimental plots resulted in a 47% decrease in leaf production of the saltmarsh grass *Spartina* in a single season (Bertness 1985). Removal experiments are less easy to carry out in mangroves, but suggest similar effects. The removal of crabs from 15 m×15 m enclosures within Australian mangroves resulted in a significant increase in soil sulphide and ammonium levels, and a reduction in cumulative growth and reproductive output of the trees (Smith *et al.* 1991). Burrowing crustaceans are, therefore, significant ecosystem engineers. Not only do they alter the physical structure of their environment, but they also significantly affect the growth and productivity of the mangrove trees.

6.3.2 Molluscs

6.3.2.1 *Snails*

The most conspicuous mollusc inhabitants of mangroves are gastropods. As with crustaceans, these include deposit feeders, herbivores which browse on fallen leaves or on algae growing on tree bark, and predators. Although a few species are unique to mangroves, most snails that live on mangrove mud are also found on open mudflats. Species that live on trees are generally also found on other hard surfaces, such as rocks. Very few are exclusive to mangroves.

The principal predatory gastropods are species of *Thais*, found in mangroves worldwide. These cruise over mud and mangrove roots, feeding on barnacles or small gastropods. *Thais* drills through the shells of its prey with the radula, a ribbon-like tongue covered in numerous sharp teeth, which operates rather like a miniature chainsaw.

Most of the snails found on the mud are deposit feeders, using their radula to scrape organic particles from the surface. Some genera, such as *Cerithidea*, have representatives in mangroves throughout the world, others are geographically more restricted (Plaziat 1984). Among the more spectacular are species of *Telescopium* and *Terebralia*, widespread in the Indo-Pacific region. The cone-shaped shells may reach 190 mm in length. Snails of this size are probably virtually exempt from predation. Crushed *Telescopium* shells are sometimes found scattered around burrows of the large portunid crab *Scylla*, but no other invertebrate predator could cope. The large size of this species has attracted human attention, and in many places it is over-exploited, as food or for other purposes. In the Vellar estuary, India, several tonnes of *Telescopium* are collected annually as a source of calcium carbonate for the lime industry (Houbrick 1991).

Few gastropods eat mangrove leaf material directly. One which does is *Terebralia palustris*. This species changes its diet with age. Young *T. palustris* are detritus feeders. On reaching a shell length of around 30 mm, they begin to cluster on fallen mangrove leaves. In small specimens, the radula carries denticulate teeth, virtually identical in size and shape to the specialized spoon-shaped setae of fiddler crab mouthparts (Figure 6.9). This remarkable example of convergent evolution reflects a close similarity in function: in both cases, the structure is used to separate fine organic particles from inorganic matter. As *Terebralia* grows, the radula teeth metamorphose to the adult pattern, becoming relatively larger, and with simple blade-like cutting edges. Examination of gut contents shows that pieces of leaf material are ingested.

Further evidence that large *Terebralia* feed almost exclusively on mangrove leaves comes from studies of the isotopic composition of the snails. Adult *Terebralia palustris* tissues show $\delta^{13}C$ values lower than $-21‰$, close to those of mangrove leaves and in contrast to the values found in species known to have other diets: the deposit-feeding fiddler crab *U. lactea*, for instance, has a $\delta^{13}C$ value of -18.96. This is good evidence that adult *L. palustris* feeds primarily on leaf material, and can be a major means of its removal from the forest floor. Juveniles show values in the range -17.5 to $-18.5‰$, and show a transition to adult $\delta^{13}C$ values which correlates with metamorphosis of the radula (Figure 6.15; Marguillier *et al.* 1997; Slim *et al.* 1997).

As *Terebralia* are often very abundant—densities of 150 adults and 475 juveniles per square metre have been recorded—they can be one of the major removers of leaf litter, particularly when sesarmid crabs are not present in large numbers (Fratini *et al.* 2004).

The most abundant snails on mangrove trees are species of *Littoraria*, close relatives of the periwinkles of temperate rocky shores. In Central American and Caribbean mangroves, on both sides of the Isthmus of Panama, and in West Africa, the common species is *Littoraria angulifera*. In the Indo-Pacific, this species is replaced by a number of others, which partition between them the slightly different habitats available on the surface of mangrove trees. In Papua New Guinea, *Littoraria scabra* and *Littoraria intermedia* are found only on the bark of trunks, roots, and branches. *L. scabra* is more common on trees at the seaward edge of the mangrove forest, whereas *L. intermedia* prefers trees facing onto freshwater creeks.

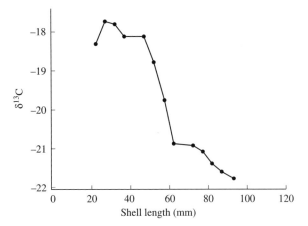

Fig. 6.15 Change in $\delta^{13}C$ of tissue of *Terebralia palustris* with increasing shell length, indicating a dietary change from detritus to mangrove leaves. Reprinted from Slim, F.J., Hemminga, M.A., Ochieng, C., Jannink, N.T., Cocheret de la Morinière, E., and van der Velde, G. 1997. Leaf litter removal by the snail *Terebralia palustris* (Linnaeus) and sesarmid crabs in an East African mangrove forest (Gazi Bay, Kenya). *Journal of Experimental Marine Biology and Ecology* **215**: 35–48. ©1997, with permission from Elsevier.

The shell colour of these two species closely matches that of the bark on which they live (as does that of other bark-living snail species). The mouth of the shell is rounded, to fit the contours of root surfaces, and the radula relatively long as an adaptation to the rough surface of the bark.

The leaves are occupied by a different species, *Littoraria pallescens*. This species has a flat aperture to match its substrate, and a much thinner and weaker shell. Shell-crushing crabs forage on roots and even trunks of mangrove trees, but not on leaves. Leaf-living snails have less need of protection against mechanical damage by crabs, and can afford to have thinner shells. *L. pallescens* is conspicuously polymorphic. Typically, up to 9% are orange or pink, 10–50% yellow, and the remainder dark in shell colour.

Stable polymorphisms such as this often result from a balance between conflicting selection pressures. Some colour varieties may be more conspicuous than others to a predator such as a bird, or predators may prefer whichever variety is the most frequent in the population. The physical environment may also affect polymorphism: yellow *L. pallescens* gain less heat from sunlight than the dark forms, and are typically around 1.5°C cooler, with the orange form intermediate. The polymorphism presumably results from a balance between such physical and biotic factors (Cook *et al.* 1985; Cook 1986; Cook and Freeman 1986).

Like crabs, mangrove snails have to survive varying salinity, high temperature, and anoxia, and avoid desiccation. Of these physical problems, salinity may be the least serious. Most mangrove snails are probably quite tolerant of a range of salinities: *T. palustris*, for example, can live comfortably in salinities of 15–35‰ (Houbrick 1991). The connected issues of heat and water conservation are probably more limiting. There is little direct information, but the clustering of many species of snail in shady positions among mangrove roots suggests the avoidance of heat stress and accompanying water loss.

Like crabs, most mangrove gastropods feed at low tide, and are therefore adapted to aerial respiration. In many species the gills are reduced (*T. palustris* being a surprising exception) and gas exchange takes place over the folded and vascularized epithelium of the mantle cavity (Plaziat 1984). High-shore species, in areas that dry out for long periods, tend to become inactive to conserve water. Closing tightly to minimize evaporation also limits gas exchange, necessitating periods of anaerobic respiration. This creates acidic conditions. Some species, particularly those found in habitats rich in detritus, appear to counter internal low pH conditions by the dissolution of calcium carbonate from their shells. Often this results in the thinning of the apex of the shell to the point where it is completely lost. This process, known as decollation, is characteristic of south-east Asian species such as *Cerithidea*, but has not been reported from comparable species in other regions (Vermeij 1974).

The external environment, of course, also tends to be anoxic and of low pH, and exterior shell erosion can be a problem. The shells of epifaunal molluscs are protected by a thick outer non-calcareous periostracum: if this is scratched, erosion of the underlying shell can occur (Plaziat 1984).

6.3.2.2 Bivalves

The most visible bivalve molluscs of mangroves are the encrusting oysters and mussels found attached to roots. Within the mud, burrowing filter-feeding bivalves may be abundant. Infaunal species tend to be found also in adjacent open mudflats, so are not peculiar to mangroves. Stable isotope analysis indicates that they do not depend directly for food on particles of mangrove detritus, but either on algal cells or on organic particles processed by other organisms.

Bivalve molluscs are also important components of the community that lives within mangrove wood, particularly in dead wood. There is some evidence that wood-boring animals can colonize wood only after it has first been attacked by fungi. Typical wood-boring molluscs are the shipworms of the family Teredinidae. (Despite their common name, shipworms are bivalve molluscs, not worms.) These include the giant mangrove shipworm *Dicyathifer*, which may reach 2 m in length; most are of more normal proportions (Plaziat 1984).

6.4 Meiofauna

Most studies of mangrove fauna consider animals visible to the naked eye, whether on the mud surface or beneath it. Like other soft sediments, mangrove mud also contains an abundant and diverse meiofauna, comprising animals small enough to live between soil particles. Figure 6.16 shows some representative meiofaunal organisms. The few studies of mangrove meiofauna that have so far been carried out have shown a little of its interest and importance.

Meiofauna are not easy to study. The organisms, besides being tiny, are often transparent and are easily damaged by being physically extracted from the mud. Extraction involves dispersing a mud sample in a suitable liquid and adding a stain such as Rose Bengal to make visible some of the more translucent animals. Then, because mangrove swamps usually are some way away from laboratories with the necessary microscopes, the samples are preserved with formalin for later analysis. Usually, this consists of separating out meiofaunal animals by sieving; meiofauna being defined, to some extent arbitrarily, by the mesh size of available sieves. In one study, meiofauna were regarded as those organisms that passed through a mesh of 1000 μm, but were retained by a mesh of 53 μm (Sasekumar 1994). Some soft-bodied organisms, such as minute turbellarian flatworms, disintegrate

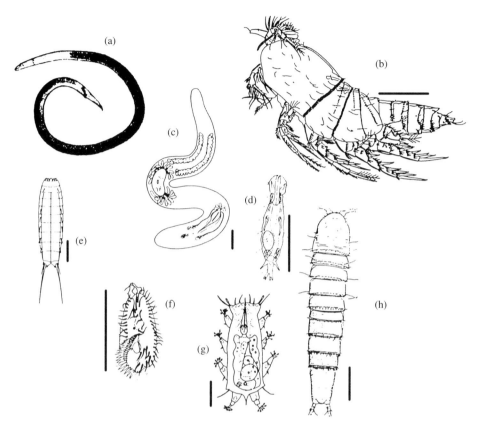

Fig. 6.16 Representative meiofaunal organisms. (a) Nematode (total body length, approx. 1 mm); (b) a harpacticoid copepod (*Prionos ornata*; scale bar, 100 μm); (c) turbellarian (scale bar, 200 μm); (d) gastrotrich (scale bar, 100 μm); (e) kinorhynch (scale bar, 100 μm); (f) ciliate protozoan (scale bar, 100 μm); (g) tardigrade (scale bar, 50 μm); (h) harpacticoid copepod (*Lizashtonia hirsutosoma*; scale bar, 100 μm). Animals (b) and (h) come from mangrove sediments in western peninsular Malaysia, and are reproduced by courtesy of Rony Huys and Mike Gee. (c)–(g) are from Higgins, R.P. and Thiel, H. (eds) 1988. *Introduction to the Study of Meiofauna*. Smithsonian Institution, Washington DC. ©1988 The Smithsonian Institution. Used by permission of the publisher.

during this process, so are underestimated when samples are analysed; and differences in the methods used by different investigators make it difficult to compare results from different sites. Nevertheless, some interesting patterns are emerging.

As a general rule, small animals are more numerous than large. It is no surprise, then, that beneath a 10 cm^2 area of mud there may be many thousands of individuals. Most of these live in the upper centimetre or so of the mud, with numbers falling off rapidly below this depth (Somerfield *et al.* 1998). Meiofaunal populations are also more diverse than the macrofauna: not only are there many species, but the level of taxonomic diversity is higher. The macrofauna is dominated by two phyla, molluscs and arthropods, with only a few other phyla being represented, while the meiofauna includes representatives of many phyla. One set of meiofaunal samples from the mangroves of Hinchinbrook Island, Australia, for example, yielded 1600 turbellarian flatworms, 200 nematodes, nine harpacticoid copepods, and sundry ciliate protozoa, foraminifera, bivalve molluscs, oligochaetes, polychaetes, hydrozoa, archiannelids, kinorhynchs, amphipods, cumaceans and other crustacea, tardigrades, and gastrotrichs (Alongi 1987a).

The abundance and composition of the meiofauna vary with the nature of the soil, with sediment depth, with shore level, with the dominant tree species, and seasonally. In Kenya, the greatest abundance was in the *Bruguiera* zone, followed by *Rhizophora*, *Avicennia*, *Sonneratia*, and *Ceriops* (Vanhove *et al.* 1992). In contrast, in Malaysia relative densities were reversed: the greatest abundance was in the *Avicennia* zone, followed by *Rhizophora*, then *Bruguiera* (Sasekumar 1994). It is not clear whether procedural differences explain such disparities, or whether genuine geographical differences are involved.

It is quite possible that meiofaunal populations also show variation at much smaller scales. If macrofaunal animals vary in relative abundance over horizontal distances of the order of a few metres, meiofaunal animals may vary in abundance over distances of the order of a few centimetres. Sampling methods used so far would have no chance of picking up variation on this scale.

Little is known of the food chains or community structure of the meiofauna. As with the macrofauna, the meiofauna operates at a number of trophic levels. There are meiofaunal herbivores, predators, and detritivores just as there are at much greater body sizes. Heterotrophic bacteria are probably a major food resource. Rates of bacterial production, at least in the Hinchinbrook Island mangroves mentioned above, are typically much greater than the maximal ingestion rates of the meiofauna, so the primary consumers among the meiofauna are not limited by food availability. High tannin levels leaching out from mangrove litter may depress meiofaunal abundance (Alongi 1987b). Variation in the tannin content in the leaves of different species may explain some of the initial differences in the meiofauna colonizing newly fallen mangrove leaves. As decay proceeds, the meiofaunal community undergoes a series of successional changes, reaching a climax state after a few weeks which is little affected by the species of leaf. The dominant organisms at this stage are nematodes and copepods (Gee and Somerfield 1997).

Does the meiofaunal microcosm interact significantly with macrobenthic organisms whose world is on such a different scale? The meiofaunal foodweb seems relatively independent, with only minor energy fluxes to the larger epibenthos. Interactions between meiofauna and the benthic macrofauna have to some extent been clarified by experiments in which the latter—particularly crabs (*Uca*) and molluscs such as *Terebralia* and *Cerithidea*—are excluded by cages from areas of mangrove mud. The results depend on area, sediment type, time of exclusion, and other factors, but in general the consequence is an increase in abundance of meiofauna.

This could be due to the elimination of predators allowing prey numbers to increase. However, predation of meiofauna by macrofauna seems to be relatively unimportant. Although fiddler crabs process a high proportion of the surface layers of sediment, they selectively remove bacteria and ciliates, and other organisms are mostly returned to the mud in pseudofaecal pellets (see p. 115). Harpacticoid copepods are eaten by juvenile fish and crabs, so might be expected to increase in numbers during an exclusion experiment. Unlike other groups, copepod abundance actually tends to decrease in exclusion experiments, and in equivalent laboratory microcosms (Ólafsson and Ndaro 1997; Schrijvers and Vincx 1997).

The meiofaunal groups which show the greatest increase in abundance in long-term exclusion experiments (several months to a year) are nematodes and oligochaetes. The species in which the increase in numbers was most marked were those known to feed on microalgae or fine, bacteria-rich organic matter. Neither nematodes nor oligochaetes are significantly eaten by crabs, so the increase in numbers is not likely to be due to relaxation of predation pressure. Despite the disparity in size, these minute organisms compete for food with the much larger epibenthic macrofauna (Dye and Lasiak 1986; Schrijvers and Vincx 1997; Schrijvers *et al.* 1997).

6.5 Fish

The creeks, pools, and inlets of mangrove forests typically have a rich fauna of fish. With such mobile animals, abundance is impossible to measure directly, although mangrove fish are often the basis of thriving local fisheries (see p. 182). Species richness is generally high. In the mangrove creeks of Selangor, western Malaysia, one survey recorded 119 species. In terms of individual numbers, 70% of the sample comprised schooling fish such as anchovies; most of the biomass was made up of larger catfish, mullets, and archerfish. Almost all were juveniles, suggesting that the mangrove habitat serves as a nursery area for the surrounding waters. The mullets (*Liza* spp.) consume significant quantities of mangrove detritus; other species generally eat mangrove invertebrates including crabs and their planktonic larvae, gastropod and bivalve molluscs, and sipunculids (Sasekumar *et al.* 1992).

How typical are the mangroves of Selangor? Figures from other mangrove areas include 83 species in Kenya, 55 and 133 from two areas of Queensland, 59 species in Puerto Rico, and 128 from the Philippines

Fig. 6.17 Mudskipper on a pneumatophore of *Avicennia*. Photograph: Hunting Aquatic Resources.

(Chong *et al.* 1990). Estimates of diversity depend heavily on catching methods and intensity, so these figures are not directly comparable. However, it can probably be concluded that mangrove creeks generally have a rich fish fauna.

Fish caught in creeks presumably forage among the mangrove trees at high tide. At low tide, this habitat is occupied by the mudskippers, relatives of the gobies (Figure 6.17). Anyone who has visited mangroves or tropical mudflats in the Indo-Pacific region will have been captivated by these fish. As their common name suggests, they skip across the surface of the mud by using their tail, and their pectoral fins which serve almost as legs. Some even climb up aerial roots, where they cling on partly by grasping the root with their pectoral fins, partly because in many species the pelvic fins are fused together to form a sucker. As the tide rises, mudskippers commute upshore (or up trees), and as it falls they move back down across the mud. Despite this mobility, they are generally territorial, sometimes maintaining high-shore burrows as well as low-shore feeding territories, in the core of which is a further burrow system. Territories are defended against conspecifics and other intruders such as crabs, and are the basis of courtship and other social interactions. Social displays involve a complex series of postures, erection of the conspicuous dorsal fin, undulations of the tail, head-butting, and biting (Macnae 1968; Clayton 1993).

Burrows vary in structure, but usually include a more or less vertical shaft with horizontal side tunnels. Sometimes the burrow is shared with a commensal pistol shrimp (*Alpheus*) or crab. Regular commensal associations between gobioid fish and alpheid shrimps are well known in other habitats, such as sandy shores; some of the other associations may be more fortuitous, resulting from the accidental anastomosis of mudskipper and crab burrows. A mudskipper burrow may have surface embellishments, such as carefully constructed mud towers or rimmed saucer-like depressions. These might have a display or other social function, although a tower at one end of a U-shaped burrow would also promote water movement through the burrow as a horizontal tidal current washed past.

Mudskippers of the genera *Periophthalmus* and *Periophthalmodon* are predominantly carnivorous and, like many carnivores, often unselective in their prey. The diet of *Periophthalmus* includes crabs, insects (particularly ants), spiders, nemertine worms, shrimps, copepods, amphipods, and snails. *Periophthalmodon* appears to specialize in crabs, particularly sesarmids and fiddler crabs. In contrast, *Boleophthalmus* is largely a deposit feeder, skimming sediment from the mud surface with a peculiar sideways movement of its head. It then moves the mouthful of food and water around the buccal cavity, possibly sorting it into its constituents in a similar way to *Uca* (p. 115), before squirting water out through its mouth and gill apertures. Other species, such as *Scartelaos*, have a more omnivorous diet.

All mudskippers are probably to some extent omnivorous when the situation demands, and there are certainly differences in diet between small and large individuals of the same species. The differences in the predominant diet, however, are sufficiently consistent within species to be reflected in morphological adaptations of the gut. As a general rule, plant material is less easily digested and herbivores therefore tend to have longer guts. This is shown by the relative gut lengths of mudskipper species, shown in Table 6.4.

Mudskippers have appropriate physiological adaptations to the demanding amphibious way of life. As long as they can return periodically to burrows, water loss is probably not a major problem. Some species, however, do have considerable ability to survive desiccation where necessary. Although it rarely spends long periods of time out of water in normal circumstances, the Chinese mudskipper *Periophthalmus cantonensis* can survive 2.5 days

Table 6.4 Relative gut lengths of mudskipper species. Because guts are difficult to measure, and vary in length, methods of measurement differ: relative gut lengths are strictly comparable only within studies. Each column represents a separate study. Data quoted in Clayton (1993).

Species	Feeding mode	Relative gut length				
Periophthalmus	Carnivore	0.5–0.6	0.64–0.76	0.39	0.28–0.45	
Periophthalmodon	Carnivore	1.09	0.43	0.8	0.42	
Scartelaos	Omnivore	1.27			0.6	
Boleophthalmus	Herbivore	2.55	1.45			
Apocrytes	Herbivore					2.0
Pseudoapocrytes	Herbivore					2.7

out of water, losing water much more slowly than a frog of comparable size. The East African *Periophthalmus sobrinus*, which spends about 90% of its time out of water, can survive water loss equivalent to one-fifth of its body mass (Gordon *et al.* 1978).

The principal nitrogenous waste of aquatic animals is often ammonia, which is toxic but highly soluble and therefore easily eliminated. Mudskippers have various ways of minimizing ammonia accumulation while out of water, including producing alternative, less toxic, nitrogenous compounds and active secretion of ammonia across the gill surface (Randall and Tsui 2002).

Respiration in air also requires adaptation. Gills are less effective as a mode of gas exchange in air than in water. It was formerly thought that mudskippers survived in air by taking water into the mouth and gill chambers to keep the gills immersed, and that when the oxygen was depleted or the water lost they had to return to the water to stock up again. It now appears that this is not the case, and that mudskippers are, to a greater or lesser extent, air breathers. *Periophthalmodon* transports air into its burrows and stores it in a dome-like chamber, where it is available as a reservoir of oxygen. This may be particularly important to embryonic development, as mudskippers commonly deposit their eggs on the roofs of their burrows. In several other species, including *Periophthalmus*, *Scartelaos*, and *Boleophthalmus*, air is present in the spawning chambers (Ishimatsu *et al.* 1998).

Gas exchange at the gills is supplemented by other parts of the body, particularly the buccal and pharyngeal epithelium and skin. The skin, in particular, is highly vascularized. In *Periophthalmus* the most important areas for respiration are the front of the head and gill covers. Blood vessels rise from the dermis and branch in the epidermis, forming umbrella-like structures parallel to the surface. Other species have vascularized papillae at the skin surface. The relative importance to respiration of the reduced gills and vascularized skin is reflected in the low values of the gill/skin area ratio for species that are more dependent on aerial respiration. In *Periophthalmus* this ratio is 0.27–0.46 and in *Periophthalmodon* 0.22–0.5. This compares with a ratio of 0.67–0.77 in *Boleophthalmus* and 0.72 in *Scartelaos*, both species believed to depend less on aerial respiration (Clayton 1993).

7 Seagrass communities

Seagrasses, like mangroves, are a major source of photosynthetic primary production, providing the energy base for an often complex ecosystem, through detrital decomposition routes as well as herbivory. The rhizomal root system stabilizes the sediment, while densely growing leaves reduce current velocity and encourage the settling of further particles from suspension. Small invertebrates and fish are also provided with protection against waves, currents, and predators. With few exceptions, the organisms comprising seagrass communities are not restricted to seagrass habitats, but are simply more abundant there than in the surrounding environment. Seagrass communities comprise epiphytes and sessile epifauna growing on seagrass leaves and rhizomes, as well as more mobile animals, many of which move between seagrass and other habitats at different times or life-cycle stages. These include herbivores (ranging from tiny feeders on microscopic epiphytic algae to massive turtles and seacows; Figure 7.1), detritus feeders, and predators.

7.1 Epiphytes

Seagrass leaves often acquire a luxurious and diverse growth of bacteria, fungi, and algae, ranging from single cells to thalli a few centimetres long, as well as a range of sessile animals. As seagrass leaves extend continuously, the epiphytic community varies with position, the oldest regions of the leaf carrying the greatest biomass and highest diversity of epiphytes. The accumulation of epiphytic biomass on seagrass leaves varies greatly, but can be considerable, and epiphyte productivity can be as much as 60% of the above-ground productivity of the seagrass itself. Epiphytes may be much smaller than seagrasses, but their rapid turnover means a contribution to productivity disproportionate to their mass (Hemminga and Duarte 2000).

A heavy epiphyte load has a detrimental effect on seagrass plants. Hydrodynamic drag may be greater, increasing the risk of leaf loss from

138 THE BIOLOGY OF MANGROVES AND SEAGRASSES

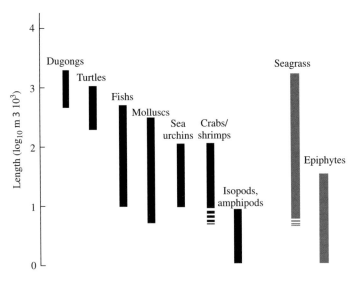

Fig. 7.1 Relative sizes of major herbivores of seagrass meadows. Note the logarithmic scale.

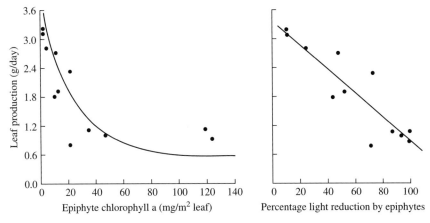

Fig. 7.2 Effects of epiphyte cover and consequent light reduction on seagrass leaf production. Epiphyte cover is measured as the concentration of chlorophyll a. From Silberstein, K. et al. (1986). The loss of seagrass in Cockburn Sound, Western Australia. III. The effect of epiphytes on production of *Posidonia australis* Hook. F. Aquatic Botany **24**, 355–37, with permission from Elsevier.

wave or current action. The principal effect, however, is on seagrass photosynthesis. Algae are efficient light harvesters, reducing the amount of light reaching the leaf for photosynthesis (although possibly also filtering out potentially harmful ultraviolet radiation). Epiphyte cover may also restrict seagrass uptake of CO_2 and mineral nutrients from the surrounding water. Excessive epiphytic growth—for example in eutrophic conditions—restricts light, reduces productivity, and may even cause mortality (Figure 7.2; Silberstein *et al.* 1986; Heck and Valentine 2006).

7.2 Molluscs

Bivalve molluscs—particularly mussels and clams—may be quite abundant within seagrass meadows, with some attached to rhizomes or leaves, and the majority burrowing in the sediment, where they may be protected against excavating predators by the rhizome mat. Bivalves are primarily filter feeders, extracting organic particles from suspension. A seagrass community generates considerable amounts of organic particulate matter through the decomposition of seagrass leaves and the production of animal faeces, whereas the reduction of water velocity by the plants (Figure 4.7) encourages the settling of fine organic particles, hence making them available to bottom-living bivalves. By removing particles from the water column, bivalves (and other filter feeders) retain nutrients within the seagrass meadow and enhance seagrass productivity.

One bivalve mollusc, the giant fan mussel *Pinna nobilis*, is virtually restricted to Mediterranean *Posidonia* meadows. *Pinna* may reach 86 cm in length, and shows the fastest recorded growth rate of any bivalve—up to 1 mm/day, a testimony to the rich food supply available in *Posidonia* meadows. About 80% of the animal's length stands clear of the sediment; because of the phenomenal growth rate, the shell is rather weak, so *Pinna* probably depends on the seagrass cover to hide from large predators (Hemminga and Duarte 2000).

Gastropod molluscs may be herbivores, detritus feeders, or predators. Many species browse the epiphytic flora on seagrass leaves, rather than the less nutritious (and in some cases possibly aversive) leaves themselves. Reduction of the impeding epiphytic layer benefits the seagrass plant, increasing productivity and reducing leaf mortality, and control of epiphytes by grazing herbivores may be a key factor regulating seagrass growth (Heck and Valentine 2006). Although for the most part gastropods feed on epiphytic algae, they may also damage or destroy the underlying epidermis of the seagrass plant itself. The limpet *Tectura depicta* can have a devastating effect on *Zostera marina*, since the epidermis contains almost all of the plant's chloroplasts, even reducing a 30-ha seagrass meadow in Monterey Bay, California, by about half (Zimmerman *et al.* 1996).

Gastropods also feed on detritus. Fine detrital particles adhere to the epiphyte layer on seagrass leaves and are ingested by grazing herbivores along with diatoms and microalgae. On a larger scale, the massive queen conch (*Strombus gigas*) has been shown, in a Bahamian seagrass meadow, to ingest large amounts of seagrass detritus, and to have a major impact also on the abundance of other animals, particularly small crustaceans, that depend on this resource (Stoner *et al.* 1995).

7.3 Crustaceans

Seagrass meadows have a diverse crustacean fauna, including amphipods, isopods, shrimps, crabs, copepods, and ostracods. These include herbivores, detritus feeders, and predators. Collectively, they are particularly important in energy flow from microalgae, other primary producers, and detritus through higher trophic levels. Seagrass meadows may also provide nursery habitats for juvenile crabs and penaeid shrimps, which spend their adult lives elsewhere.

Smaller crustaceans such as amphipods graze smaller epiphytes, often selectively. This may result in a reduction in epiphyte biomass, but an increase in biodiversity (Jernakoff and Nielsen 1997). As with epiphyte-feeding molluscs, this grooming of the fronds may benefit seagrass plants.

As in mangroves, some crustaceans consume living seagrass tissue. A large proportion of the diet of the intertidal ocypodid crab *Macrophthalmus hirtipes* in New Zealand consists of *Zostera* root and rhizome material (Woods and Schiel 1997). Most consumption is, however, of leaf material. Crabs and shrimps harvest leaf material, often taking it into burrows for later consumption. Thalassinid and alpheid shrimps take quite considerable amounts: in Indonesia, the shrimp *Alpheus edamensis* has been shown to account for more than half of total leaf production (Stapel 1997). Leaf-mining isopods burrow through *Posidonia* tissues: although minute, they may cause significant secondary damage by allowing the entry of water and pathogens into the tissues, or by weakening leaves and increasing the chances of leaf loss due to wave action (Gambi *et al.* 2003).

Thalassinid shrimps also demonstrate another parallel with mangroves: as ecosystem engineers (p. 125). Their burrowing activities disturb the sediment, so that the succession of species from good colonizers of disturbed substrate through to dominance by successful competitors is disrupted. Typical patch sizes cleared by shrimp activity are on a scale of around 0.5 m, with a density of about 3 mounds/m^2. This implies a complete reworking of the sediment every second year or so, and overall disturbance probably greater than that of the more obvious but much less frequent typhoons. This probably explains the frequency of mixed-species meadows in areas such as the Philippines, contrasting with the more general monospecific pattern elsewhere (Hemminga and Duarte 2000).

7.4 Echinoderms

Sea urchins can be extremely abundant in seagrass beds, where they eat epiphytes, fresh leaves, detritus, or a combination of these. The purple urchin *Lytechinus variegatus* in the Gulf of Mexico can completely strip a seagrass bed of leaves, leaving extensive bare patches. The local impact of urchins on seagrasses fluctuates enormously. One interpretation is that young urchins are more easily caught by predators in bare areas, leading to a fall in urchin numbers. Relieved of the presence of their dominant herbivore, seagrasses recover, leading in turn to accumulation once again of high densities of urchins. Interactions between seagrasses and urchins cause populations of both to oscillate, over a period of (probably) years (Heck and Valentine 1995). Similar fluctuations have been observed in other parts of the world, for example in the Philippines, where two species of urchin coexist in large numbers in seagrass meadows. *Tripneustes gratilla* and *Salmacis sphaeroides* both eat fresh seagrass leaves, particularly *Thalassia hemprichi*. The diet of *Tripneustes* is dominated by *Thalassia* (77–89%), but *Salmacis* also consumes seagrass detritus and algae (up to 39 and 30%, respectively) as well as other items. Presumably this reduces competition between the two species. At different times of year, urchin densities vary from 0.9 to 4.2 m^{-2}, and overall impact of urchin grazing varies from less than 5% to more than 100% of seagrass production (Klumpp *et al.* 1993).

The other important echinoderms in seagrass are the holothurian sea cucumbers, which ingest sediment and extract seagrass detritus and other organic matter from it. Their burrowing activity also affects the sediment, which may benefit seagrasses by rendering it more suitable for colonization. Species found in seagrasses include the commercially important bêche-de-mer or trepang (*Holothuria* spp.) and *Synapta maculata*, a snake-like species that reaches a staggering 3 m in length.

7.5 Fish

The fish population of a seagrass meadow may be highly diverse. Some species are permanent residents, and others live in seagrasses only at certain stages in their life cycle (for example, as juveniles) or are seasonal occupants; many commute in from adjacent habitats. Because of these movements in and out of seagrasses, fish are probably one of the main routes of energy flow linking seagrass meadows with other ecosystems.

Fish are attracted to seagrasses for protection against predators, as well as for the food resources offered. In some species, taking refuge in seagrass meadows means relinquishing a better food source elsewhere. The naked goby (*Gobiosoma bosci*) of the temperate west Atlantic is one species that has been shown to trade off food against security: it lives in *Z. marina* beds, but grows more slowly there than on open sand (Sogard 1992).

Of those fish which depend on seagrass for food most are predators, exploiting the copious invertebrates, or smaller fish. Others are omnivorous, detritivores, or herbivores. Most herbivores graze on epiphytic algae rather than on the seagrass itself, for example the Mediterranean *Sarpa salpa* (Sparidae). Many species mix epiphytes with seagrass material in their diet, among them parrot fish (Scaridae), leatherjackets (Monacanthidae) and rabbitfish (Siganidae). Relatively few graze directly on living seagrass leaves. In the Caribbean, parrotfish (*Scarus* and *Sparisoma* spp.) make sorties from coral reefs to graze on adjacent seagrass beds, sometimes to the extent of completely clearing seagrass plants and leaving a zone of bare sand around the reef (Randall 1965). Generally, however, the proportion of seagrass leaf production eaten by fish seems to be low, although it is greater when competing sea urchins are removed (McClanahan *et al.* 1994).

Although often little living tissue is taken directly from the plant, bites out of a leaf may weaken it and cause it to break off and float away. Halfbeaks (*Hemiramphus* spp.) feed at the surface of the water, catching invertebrates and small fish, but also biting pieces out of floating leaves of *Syringodium* (in preference to *Thalassia*, where both are present; Randall 1965).

Many of the fish of seagrass beds are predators, feeding on other fish or on shrimps, amphipods, polychaete worms, or other small invertebrates. Predator–prey interactions are complex. The complexity of the seagrass environment makes more difficult the location and pursuit small and active prey; this is one reason why seagrass beds are an important nursery area for many fish species. On the other hand, ambush predators use seagrass for concealment and wait for prey to come to them. One group of specialist sit-and-wait predators are the pipefish and seahorses (Syngnathidae). Pipefish, particularly when lying vertically, look very similar to surrounding seagrass blade and mores, and are very hard to spot. Seahorses can attach themselves to the seagrass by their prehensile tail and remain motionless until prey come within striking distance.

Larger predatory fish, such as rays, may also affect the seagrass habitat indirectly. By physically disturbing the sediment, they may disturb plant succession, and favour more rapidly growing, colonizing species. Bioturbation may have similar effects on infaunal animals.

The composition of the fish fauna of seagrass beds varies greatly with time, and from one area to another, so no simple generalizations can be made. Moreover, human fishing activity must have had a great impact, and it is unlikely that many of the seagrass habitats studied are in anything like a pristine state. Fisheries are selective, and often fish down the food chain: this involves first taking large carnivorous species (hence possibly increasing the numbers of smaller predators and of herbivores) and then, as large predators become depleted, attention focuses on smaller predators, then on herbivores. It is not clear to what extent this has been a feature of seagrass-associated fisheries, or precisely how the seagrass community has altered.

7.6 Turtles

The green turtle *Chelonia mydas*, widespread in the Atlantic, Mediterranean, and Indo-Pacific, is a major consumer of seagrass. More than 90% of the stomach contents is likely to comprise seagrass, and a 66-kg turtle may consume more than 200 g (dry weight) of seagrass each day. Even after serious depletion by overfishing and other human activities, turtles are still significant players in many seagrass ecosystems.

The preferred food is *Thalassia testudinum*, known as turtlegrass, although other seagrass species, and algae, are also eaten. Turtles bite at the lower parts of *Thalassia* leaves, causing the uppermost part to break off and float away. The tissues of the base, relatively richer in protein and lower in lignin, are more nutritious. Because they are younger than the extremities, leaf bases have accumulated very little epiphytic growth: many epiphytic algae are calcified and less digestible than seagrass tissues.

Cropping stimulates growth, and young, growing leaves are more nutritious than older ones. In the Caribbean, turtles repeatedly crop clearly marked areas within a seagrass meadow, maintaining the plants in their so-called cultivation patches in an actively growing and more nutritious state. Diet is optimized by manipulating their environment. The stress of repeated cropping eventually leads to depletion of sediment nutrients and a decline in leaf production, hence diminishing returns to the cultivator. At this point—which may take more than a year—the turtle will abandon its patch and shift its gardening activities elsewhere (Bjorndal 1980; Thayer *et al.* 1984).

The chief problem with digesting plant material is that the more nutritious cell contents are surrounded by relatively indigestible cell walls of lignin and cellulose. The two practical solutions to this problem are mastication or microbial digestion. Smaller herbivores, such as sea urchins and parrotfish, adopt the former method and release cell contents by finely grinding the seagrass leaves. Turtles (and other large herbivores such as dugongs) can afford to ingest large amounts of material into a capacious gut, and allow their gut microflora to do the hard work of digestion. Approximately 90% of dietary cellulose is digested in this way (Bjorndal 1980).

Analysis of the gut contents of turtles has made it possible to elucidate the processes of digestion and absorption. Nitrogen, generally in the form of amino acids released by digestion of proteins, is a key nutrient: on average only about 34% of the total ingested nitrogen is absorbed between oesophagus and rectum, the rest being released with the faeces. Faecal material consists of finely fragmented remains of seagrass leaves, richer in nitrogen than the fresh leaves ingested. What is returned to the environment is therefore of higher nutritional quality than what was taken in. Turtle digestion achieves much the same result as detrital decomposition pathways, and by rather similar processes, but at about 70 times the rate (Thayer *et al.* 1982).

Turtles therefore exploit a rather unpromising food source by selectively feeding only on certain parts of the plant, by manipulating plant growth to enhance the nutritional value of their food, and by relying on other organisms—a resident cellulose-digesting microflora—to process the more intractable components of their diet. As they do so, they affect their environment by altering seagrass growth patterns, and by greatly accelerating the detritus cycle.

7.7 Dugongs and manatees

The only significant mammalian seagrass herbivores are the seacows: the dugong (*Dugong dugon*) of the Indo-Pacific and the manatee (*Trichecus manatus*) of the Caribbean and Gulf of Mexico. Little is known of a third species, the West African manatee *Trichecus senegalensis*. Manatees move in and out of fresh water, but the dugong is entirely marine, and feeds almost exclusively on seagrass.

Seacows are particularly vulnerable to human activities, but significant dugong populations still exist, particularly in the Persian Gulf, and in south-east Asian and Australian waters. Dugongs mostly forage singly, but in Australia they often form foraging herds of 100 or more (Preen 1995). Unlike turtles, dugongs do not just eat seagrass leaves: they rummage in the substrate and may ingest rhizomes as well. Somehow they avoid ingesting large amounts of sand: the head-shaking behaviour that has sometimes been observed during feeding may be a way of getting rid of sand before a mouthful of seagrass is swallowed. Although seagrass material may be partially broken down by the (rather rudimentary) teeth, most of the digestion is probably achieved by microbial digestion in the capacious intestine and caecum. A dugong's large intestine may be 30 m in length, and it takes several days for food to pass along it (Lanyon *et al*. 1989).

Dugongs require a food intake of up to a quarter of their body weight per day; and an adult dugong can weigh 250 kg. Typically, dugong foraging leaves meandering trails through the seagrass bed, up to a quarter of a metre wide and excavated 3–5 cm into the substrate. This appears crude, but dugongs are actually quite selective feeders. With some seagrass species, such as the more nutritious *Halophila*, the entire plant is eaten, rhizome and all. With other species (*Thalassia hemprichi*) only leaves are selected. Where a choice of species is available, the foraging trails meander through patches of the preferred species: in Thailand, dugongs feed exclusively on *Halophila ovalis*, although many other species are available (Thayer *et al*. 1984; Lanyon *et al*. 1989; Nakaoka 2005).

Foraging would be less efficient if one dugong attempted to feed in a strip that had only recently been exploited by another. Dugong feeding trails rarely overlap. This means, in practice, that patches of undisturbed seagrass are always present between trails, never more than about 1 m apart. These rapidly expand and coalesce to reoccupy the depleted area. *Halophila* grows more rapidly to occupy empty space than most other species. Dugong disturbance of the habitat therefore favours this pioneer species at the expense of more slow-growing species, such as *Zostera*, which without disturbance would prove better competitors and would eventually dominate the environment (Preen 1995).

Dugong grazing therefore promotes seagrass species diversity. The grazing strategy also ensures a continuous supply of actively growing and more nutritious *Halophila*, a species that is in turn more nutritious than alternative species. Although their grazing methods differ from those of turtles, dugongs also show a form of cultivation grazing and manipulate their environment to optimize feeding.

7.8 Birds

In temperate and subtropical regions, swans, geese, and ducks can be important consumers of seagrass growing in the intertidal zone and shallow sublittoral zone. Estimated feeding rates suggest that waterfowl can ingest seagrass at a much greater rate than other herbivores. Seagrass leaves are physically macerated in the gizzard, and pass rapidly through the gut, too rapidly for microbial digestion to be significant. Probably the assimilation efficiency is low, and most of the ingested material is returned to the environment where it enters the detrital system (Thayer *et al.* 1984; Hemminga and Duarte 2000).

Many intertidal seagrass beds provide a vital resource, at certain times of year, for migrating birds. Large areas of seagrass can be virtually eliminated. In one study in the Dutch Wadden Sea, more than 50% of the standing stock of *Zostera noltii* was removed over a 3-month period by ducks, mainly pintail (*Anas acuta*) and wigeon (*Anas penelope*). When above-ground leaves and shoots have been removed, brent geese (*Branta bernicla*) probe for rhizomes (Jacobs *et al.* 1981; Fox 1996).

On the Pacific coast of North America, in Alaska, vast numbers of migrating brent geese assemble in late August on one of the world's largest intertidal stands of *Z. marina*. After feeding there for a few weeks they undertake a non-stop 3-day flight to *Zostera* beds in California and Mexico. In the Gulf of Mexico, about 75% of the world population of the redhead duck (*Aythya americana*) depends on *Halodule wrightii*. In New Zealand, *Zostera capricorni* is a major component of the diet of the black swan *Cygnus atratus* and in Japan *Z. marina* is eaten by the whooper swan, *Cygnus cygnus* (Green and Short 2003).

Waterfowl are significant seagrass herbivores in many parts of the world, consuming large amounts of leaf and rhizome material, facilitating detrital breakdown pathways, and probably having an effect on the physical environment and community structure of seagrass habitats. As waterfowl are often migratory, their activities will also contribute to energy and material flow between seagrass and other habitats.

8 Measuring and modelling

Earlier chapters have dealt with the inhabitants of the mangal and seagrass meadows. The principal characters having been established. This chapter will consider the plot: how do members of these communities interact?

The major transactions between species can be thought of in terms of energy flow or, as energy itself is difficult to measure directly, as carbon fluxes. Nutrient cycles, particularly of phosphorus and nitrogen, are also highly important, and interact with energy flow.

8.1 Mangroves

The basis of all ecosystems is autotrophic primary production. Trees are all autotrophic primary producers. Species differ in photosynthetic rate, architecture, salinity tolerance, and in many other respects, but these differences are much less significant than the over-riding similarity between species. For practical purposes, mangrove trees can be treated as a homogeneous functional group.

In the same way, sesarmid crabs can be lumped together and treated as a further functional group. Most species subsist largely on decaying mangrove leaves. They are not very different in size, and often remarkably similar in appearance and behaviour. Again, the similarities outweigh the differences. Similarly, deposit feeders, or filter feeders, can be regarded as constituting further functional groups.

This is not to say, of course, that differences between broadly similar species do not matter. To achieve a broad quantitative understanding of a mangrove ecosystem, it is practical to treat, say, sesarmid crabs as a homogeneous functional group and ignore the differences between species, and even the number of species being lumped together. Species diversity within a functional group may well be an important variable: if 10 species of sesarmid are present, each with slightly different behaviour and food preferences, is leaf litter processed more rapidly than if the same number of crabs belong to a single species? One major current controversy in ecology is over the general relationship (if any) between biodiversity and ecosystem function. Are there any important general differences in productivity or stability between a species-rich ecosystem and a species-poor one? The issue is discussed in a later chapter.

By judiciously combining species into functional groups, and analysing their quantitative relationships, it is possible to produce a robust descriptive model of an ecosystem. Such a model could be refined almost endlessly, but there are limits to the usefulness of such fine tuning. A quantitative model is only as strong as its weakest data, and refining one set of measurements is pointless while another is known only within wide margins.

This chapter attempts to quantify some of the major carbon fluxes through some of the principal functional components of ecosystems.

8.1.1 How to measure a tree

Photosynthesis and primary production are basic elements of the mangrove ecosystem. Microalgae and cyanobacteria contribute significant photosynthesis and primary production on the surface of the mud and of the trees, as well as in mangrove creeks and adjacent waters. The principal primary producers in most mangrove ecosystems, however, are the mangrove trees themselves. Two key features that must be quantified are the biomass of mangrove trees, and their rates of photosynthesis and primary production.

8.1.2 Biomass

Biomass is the sum total of all of the components of a tree, below ground as well as above ground: aerial and underground roots, trunk, branches, leaves, flowers, and propagules. To be of value, biomass must be related to area. Theoretically, one might simply dig up a given area of mangrove habitat and weigh all of the components of all trees in that area. This heroic operation is scarcely a practical proposition: besides, it would rule out the possibility of recording changes in biomass with time.

Mangrove biomass is therefore usually measured by estimation. The most common approach is to describe the population structure by measuring individual trees, and to convert this to biomass by using an established relationship between tree size and biomass.

A common practical measure of tree size is the diameter of the trunk at breast height (DBH). An alternative measure is the circumference, or girth, at breast height (GBH): this is generally preferable, since mangrove trunks are often far from circular in cross-section and therefore do not have a single diameter. Because mangrove biologists differ in size, breast height is often taken as a standard 1.3 m. Even then, some flexibility is required in accommodating, say, *Rhizophora* trees where aerial roots spring from the trunk more than 1.3 m above the soil surface, or where a tree divides into two or more separate stems below this height (Figure 8.1). Further problems arise where mangrove trees are stunted, and a mature tree may never reach a height of 1.3 m.

Fig. 8.1 A scientist (Liz Ashton; see also Figure 6.16h, which is named after her) attempting to measure the diameter at breast height (DBH) of a *Rhizophora* tree in the Matang forest, Malaysia.

To convert DBH (or GBH) to tree biomass requires cutting down a sample of trees over a range of sizes, weighing them, and establishing the relationship between DBH and dry weight or biomass. This relationship is exponential, summarized by the general equation

$$\text{Dry weight} = A \times (\text{DBH})^B$$

where A and B are constants. It is then possible to arrive at an estimate of the biomass of a tree by measuring its DBH, or of the biomass per unit area by measuring the DBH of all trees within that area.

There are limitations on the accuracy of this method. The exponential relationship may not hold for the entire range of tree sizes. Seedlings, for example, are very different in shape from mature trees, and are not heavily lignified. It is unlikely that they will fall on exactly the same exponential curve as adult trees. This limitation is probably trivial if an estimate of biomass/area is required, as the collective contribution of seedlings to total biomass will be slight compared with that of mature trees. A greater problem is that the relationship does not hold for all members of a species under all circumstances, and may differ between sites. Finally, of course, each species will have a different relationship between biomass and DBH.

To move from a measured relationship between biomass and DBH to an estimate of biomass per unit area requires either measurement of every tree in the area, or at least knowledge of the population size structure of a sample of the trees. This can vary widely, particularly after disturbance (Figure 8.2).

Total above-ground biomass varies widely, typically being highest at low latitudes and declining northward and southwards from the equator. Some undisturbed *Rhizophora* forests in northern Australia may reach values of up to 700 t of dry weight/ha, but a range of 500–550 t/ha is more usual in old mangrove forests of south-east Asia, equivalent to perhaps 250–275 t of

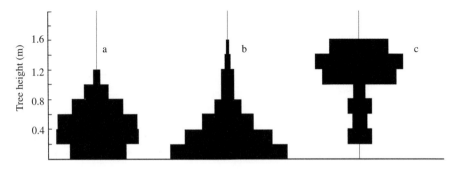

Fig. 8.2 Size structure (height in metres) of *Avicennia marina* populations at three sites near Xiamen, China. The sample sizes differed, and the horizontal bar indicates 10% of each sample.

carbon/ha. In contrast, the dwarf *Rhizophora* of Florida may have a biomass of as little as 7.9 t/ha (Clough 1992; Saenger and Snedaker 1993).

Above-ground biomass consists of leaves, branches, the principal stem of the tree and, in species such as *Rhizophora*, aerial roots. Typically, leaves contribute 3–5% of the total, branches 10–20%, and the main trunk 60–90%, while aerial roots, where they are present, may amount to between 8 and 25% (Clough and Scott 1989). Below-ground biomass is also important, but here there is little reliable information. It is difficult enough to reach a reasonably accurate figure for above-ground biomass, but the problems are compounded with below-ground components. Roots have to be dug out of the mud, and cleaned before being weighed. The tangle of absorptive roots must then somehow be separated into live and dead elements. Some reports conclude that mangroves have a high proportion of below-ground biomass compared with other tropical trees: 30–50% of the total in *Rhizophora*, for which much of the root mass is above ground, and 50–62% in *Avicennia* with its extensive, radially spreading underground cable roots (Snedaker *et al.* 1995). In contrast to these estimates, the below-ground biomass of *Rhizophora* in Malaysia has been calculated at less than 20% (Gong and Ong 1990). The ratio of below-ground to above-ground biomass varies with environmental conditions (Chapter 2).

8.1.3 Estimating production

Tree biomass is important, but it is not itself a measure of production rate or of energy flow. If biomass is measured in mass per unit area, the logical measure of productivity is dry-weight mass per unit area, per unit time. This is commonly expressed as tonnes of dry weight per hectare per year, or as grams of carbon per square metre per year. (Carbon amounts to 40–45% of tissue dry weight, and 1 g of carbon, in this context, equates to an energy of 12 cal, or 50 kJ.)

The next step is to estimate primary production: the rate at which organic matter is produced as the result of photosynthesis. Direct measurement of photosynthetic rates can be carried out, with suitable instrumentation, in a well-equipped laboratory. In the field, it is often necessary to resort to indirect and approximate methods. One widely used method entails measuring the attenuation of light as it passes through the mangrove canopy. This can be done simply by measuring incident light directly under the canopy and comparing it with light above the canopy or, in practice, the light reaching the ground in an adjacent open space. If we know the assimilation coefficient of the trees—the number of grams of carbon fixed per unit of chlorophyll—and make some simplifying assumptions about the relative amounts of chlorophyll per leaf in leaves at different levels of the canopy, it is then possible to calculate potential primary production, at least at the moment at which the measurements are taken (English *et al.* 1997).

This estimates *gross primary production*. The advantages of the method are its speed of operation and the simplicity of the field technique. The limitations are that it depends on factors such as the assimilation coefficient, which will vary with different conditions of climate and salinity, so at best what is obtained is a snapshot view. A proportion of the gross primary production is used by trees in respiration. When these running costs are deducted, what remains is *net primary production*. The rest of the ecosystem depends on this net primary production (and that contributed by other sources, such as microalgae).

Respiration rates of leaves can be estimated directly, even in the field, by measuring gas-exchange rates. Measuring the respiration rate of the woody components of the tree is less easy, and estimating respiration of the underground root systems is virtually impossible.

Most measures of net primary production have therefore relied on measuring the production of biomass directly. Net primary production should be equal to the increase in standing crop biomass, plus the biomass shed in the form of leaves, twigs, branches, and reproductive structures. If net primary production is calculated on the basis of an area of mangrove forest, rather than being related to single trees, the periodic death of whole trees should be included. If a forest is in a state of equilibrium, losses from tree death and shedding of components will balance total biomass production.

Accumulation of biomass is estimated by using the methods of biomass estimation described above, and repeating them after appropriate time intervals, preferably over at least 5–10 years. The most reliable estimates come from managed mangrove plantations, where productivity may be very different from virgin forests. In the intensively managed Matang forest of western Malaysia (see Chapter 11) the mean annual increment in above-ground biomass of *Rhizophora* was 18 t/ha. This is similar to estimated annual increments from Thailand, of 14–33 t/ha. An unmanaged section of the Matang forest, in contrast, gave a much lower estimate of 6 t/ha per year. This is low in comparison with unmanaged *Rhizophora* forests in Australia, where accumulation was estimated at 6.3–45.4 t/ha per year (Ong *et al.* 1984; Putz and Chan 1986; Clough 1992).

In comparison, measurement of litter production is conceptually straightforward, although many practical points need to be considered. Appropriate-sized mesh litter traps are positioned within the forest in such a way as to trap all material falling from the tree canopy. They must be below the portions of the trees from which material will fall, but high enough to ensure that leaves are not washed away by the tides, or eaten by crabs. The contents must be collected frequently to reduce losses from decay, and over a sufficiently long period to record seasonal and other fluctuations in litter production. Finally, the number of litter traps must be sufficient, not just to achieve statistical validity, but to offset the inevitable losses when local people think of alternative uses for the materials of which they are made.

Despite these practical problems, litterfall measurements have been widely carried out, and often used as a convenient surrogate for full determinations of net primary production. Typical figures range from 5 to 15 t/ha per year, although in dwarf mangroves production may be as low as 2.9 t/ha per year. Litterfall is often strongly seasonal, particularly outside the tropics (Figure 8.3; Steinke and Ward 1988; Saenger and Snedaker 1993).

Biomass, like litter production, is lower at higher latitudes. The latitudinal trends, however, are not in proportion. The relative amount of litter produced, or litterfall/biomass ratio, increases with latitude. It also decreases with increasing size of tree: small mangrove trees produce relatively more litter

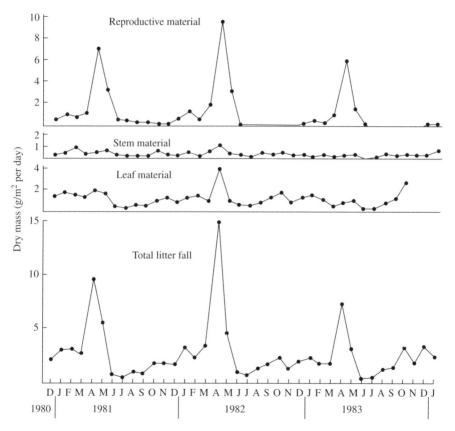

Fig. 8.3 Mean daily litter production of an *Avicennia* population in South Africa. Reproduced with permission from Steinke, T.D. and Ward, C.J. 1988. Litter production by mangroves. II. St Lucia and Richards Bay. *South African Journal of Botany* **54**: 445–454. ©South African Association of Botanists.

than larger ones, presumably because leaves make up a higher proportion of their total above-ground biomass (Saenger and Snedaker 1993).

8.1.4 What happens to mangrove production?

As soon as a fallen leaf lands on the forest floor, it is available to other organisms as an energy source. What is the fate of mangrove productivity?

In the short term, the main contribution of the trees is the continual fall of leaves, together with components of flowers, and propagules. Although the objective of every propagule is to become a tree, only a tiny minority achieve this ambition (p. 33), and the majority, together with other components of the litter, are simply an energy source for heterotrophs.

Branches occasionally drop, and even more infrequently a whole tree will fall. These irregular contributions of woody material must also be considered. The underground components of a mangrove tree also add to the available organic matter. A redundant rootlet is the subterranean equivalent of a shed leaf. In other trees, the never-ending quest for nitrate and phosphate involves production of new rootlets and the abandonment of old ones that have depleted their surroundings of these nutrients. Virtually nothing is known of the turnover of absorptive roots of mangroves, except that it presumably occurs.

The dead material shed by mangrove trees, or necromass, has four possible fates. It may rapidly be broken down, for example by sesarmid crabs (p. 103 and below); it may be decomposed by microbial action; it may be flushed out of the mangal by tides or river flow; or it may accumulate within the mangrove mud. The proportion exported will depend on prevailing conditions of tide and current. Export of material may be important to adjoining habitats, and will be discussed in Chapter 9.

The principal routing of organic material within the mangrove ecosystem is through the pathways of microbial decomposition and breakdown by macrofauna. Figure 8.4 illustrates the immediate fate of litterfall in mangrove forests in tropical Australia, and shows the major differences that depend on shore level. At lower shore levels, tidal movements rapidly flush most of the leaf litter out of the mangrove forest. At higher levels, tidal export is less. If crabs are abundant, they may consume the bulk of the litter; if not, litter tends to accumulate and microbial breakdown is more important.

8.1.4.1 *Microbial breakdown*

Left to decay, leaves decline in mass. Decay rates can most readily be assessed by placing known masses of mangrove leaves in mesh litterbags which exclude macrofaunal leaf-eaters such as gastropods and crabs, leaving them in various positions within the mangal, and weighing the

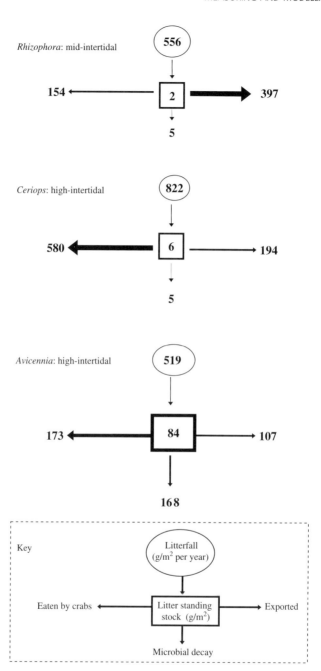

Fig. 8.4 Initial fate of leaf litter in mangrove forests in tropical Queensland, Australia. Figures in ovals and rectangles represent litterfall and litter standing stock (necromass), respectively. Data, with permission, from Robertson, A.I., Alongi, D.M., and Boto, K.G. 1992. Food chains and carbon fluxes. In *Tropical Mangrove Ecosystems*. Coastal and Estuarine Studies, no. 41 (Robertson, A.I. and Alongi, D.M., eds), pp. 293–326. Copyright by the American Geophysical Union.

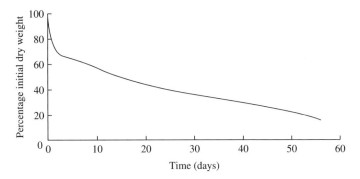

Fig. 8.5 Decline in dry weight with time of *Sonneratia* leaves in the Matang forest, western Malaysia, as percentage of initial dry weight. (Data collected by Liz Ashton.)

remains at intervals. Figure 8.5 illustrates the decline in dry weight of *Sonneratia* leaves at a mangrove site in western Malaysia.

Decomposition is more rapid in leaves immersed in mangrove creeks or adjacent waterways, or which lie in a position frequently inundated by the tides. Because the decline in weight is approximately exponential, decomposition rate can conveniently be summarized as the time taken for mass to fall to half of its original value, or half-life. The half-life of *Avicennia* leaves which are permanently immersed is 11–20 days, whereas in the intertidal zone it may be 90 days. Rate varies also with species: *Rhizophora* leaves, for instance, break down at about half the rate of *Avicennia* leaves (Robertson *et al.* 1992).

For the first 10–14 days, almost all of the loss in weight is due to physical leaching of dissolved organic carbon (sometimes referred to as dissolved organic matter) from the leaves, rather than by microbial decay. Approximately 30–50% of the organic matter of the leaves is leachable in this way, and much of what remains is insoluble structural carbohydrate such as cellulose. This is subsequently attacked by extracellular enzymes secreted by bacteria or fungi. Among the substances which leach out over the first few days are tannins (p. 107), which inhibit bacterial growth, probably by precipitating secreted enzymes and soluble substrates. Presumably the bacterial invasion takes place after tannin levels have fallen (González-Farias and Mee 1988).

After a few days, the leaf surface has been colonized by a dense bacterial population; bacterial cell densities of up to 6×10^8 cm^{-2} have been recorded on leaf surfaces after only 6 days, with a cell production rate of up to 8×10^6 cm^2/h. (Benner *et al.* 1988). Numerous species of bacteria are probably involved, although very little is known about their differing roles and interactions, or about the ecology of mangrove bacteria in general. One study of bacterial decomposition failed to isolate any bacteria capable of digesting cellulose, so it may be that bacteria are more important in processing the soluble matter that leaches out. The leached material tends to form an aggregate, which, after colonization by microbes, develops a C/N ratio of around 6 (see Figure 6.5). It is therefore a nutritious food source for small invertebrates, and is one of the main constituents of particulate organic matter (alternatively called particulate organic carbon) in the sediment (Steinke *et al.* 1990; Robertson *et al.* 1992). (The distinction between particulate and dissolved organic matter is to some extent an artificial one, since in fact there is a spectrum from truly dissolved material, through colloidal suspensions, to sizeable and discrete particles.)

Slightly later in the succession, fungi in large numbers appear on the decomposing leaf surface. The role of fungi in mangrove-leaf decomposition is also poorly understood. Many species are present; some occupying leaves soon after they reach the mud surface, others only after some time. Two genera in particular, *Trichoderma* and *Fusarium*, are powerful digesters of cellulose and play a major part in leaf breakdown (Steinke *et al.* 1990; Hyde and Lee 1995).

Bacteria and fungi actively break down leaf material, and convert much of it into their own biomass. Microbes themselves constitute a major food source for other organisms, such as deposit-feeding fiddler crabs. Even the sesarmids, which overtly feed on actual leaf material, may in fact be extracting much of their nourishment from the colonizing microbes. Meiofauna (p. 130) are also found associated with decaying leaves, and many feed on microbial cells. Like microbes, they are inhibited by tannins (Alongi 1987b).

8.1.4.2 *Crabs and snails*

Crabs and snails usually consume a significant proportion of leaf litter. How important is this to energy flow through the ecosystem?

As with trees, crab production—in this case secondary, not primary, production—must be measured. The problem is essentially similar to the measurement of net primary production of trees, where some ingenuity is needed to assess biomass increase and litterfall. The biomass of sesarmid crabs per unit area is analogous to the standing crop biomass of trees. Crabs are distinctly more mobile than trees, and can hide down burrows: this makes biomass estimation difficult, but not impossible. The parallel to leaf litterfall is the combined mass of crab faeces, moultskins, and larval release. This is measured, not by litter traps, but by estimates based on laboratory observations of faecal production and moulting frequency, and knowledge of egg production rates and larval weights. Crab mortality to predators must also be assessed, since if the crab population is roughly in equilibrium this harvest will account for the major part of crab production.

Given the difficulties, it is not surprising that there have been few estimates of sesarmid crab productivity. On one Malaysian mangrove shore, the two major species of sesarmid, *Perisesarma onychophora* and *Episesarma versicolor*, were present at average densities of 4 and 0.3 individuals/m^2, corresponding to biomasses of 2.56 and 1.95 g/m^2. Annual production was estimated at 0.7 and 9.1 g/m^2, respectively (Macintosh 1984).

Leaf-eating crabs are important for energy and carbon flow through the mangrove ecosystem in several distinct ways. By converting leaf into crab material they provide a food source for predators such as fish and birds—and, of course, other crabs. They influence energy flow in two ways other than by assimilating leaf material directly. In the first place, as described in Chapter 6, they remove a high proportion of the leaf litter into burrows. Whether or not it is then eaten, the leaf material is retained within the mangal and not flushed out by the tide or currents.

Finally, and possibly most importantly, a high proportion of leaf material is returned to the environment in the form of faeces. Crab faeces are richer in nitrogen than uneaten leaf material: passage through a crab gut enhances the nutritional quality of mangrove leaves (Lee 1997). The hindgut contents of a sesarmid often consist almost entirely of small leaf fragments, often similar in appearance to those in the foregut. A key role of the crabs is to break down the tough leaf material into smaller fragments. This accelerates subsequent microbial breakdown by increasing the available surface area, and also generates smaller particles suitable for deposit feeders. In one Australian *Bruguiera* forest, sesarmid crabs produce faeces equivalent to around 260 g of carbon/m^2 per year (Robertson and Daniel 1989). Overall, crabs accelerate litter breakdown by up to two orders of magnitude (Robertson *et al*. 1992). From an ecosystem perspective, sesarmids are more important as food processors than as assimilators.

8.1.4.3 *Wood*

Woody tissue makes up 20–50% of the total net primary production of mangrove forests, so its fate is important to understanding energy and carbon flow through the ecosystem. Wood is broken down much more slowly than leaf litter. Lignocellulose, which makes up much of the bulk of woody material, is highly refractory to decay. Trunks and large branches disintegrate over a timescale of a few years, compared with weeks or months for leaves. Small twigs break down more rapidly.

The initial breakdown of wood is largely accomplished by animal activities. If a mangrove tree dies standing, termite activity rapidly removes much of the woody material. Fallen branches and trees become honeycombed by burrows of teredinid molluscs (p. 130), aided by cellulose-digesting bacterial symbionts. In tropical Australia, it has been observed that within 4 years about 40% of the cross-sectional area of a fallen *Rhizophora* trunk has been replaced by shipworm tubes, accounting for about 90% of the loss in wood mass over that period. Half of the original mass is lost within 2 years where teredinids are present, and only about 5% where they are absent (Robertson

1991). Along with shipworm burrowing, microbial decay of the remaining material takes place, particularly through the actions of fungi.

Fibrous roots decay more slowly than the major radial roots. In 270 days, fibrous roots of *Avicennia* lost only 15% of their mass, compared with 60% for the major roots (van der Valk and Attiwill 1984). This probably reflects the enormous surface area within the major roots, on which bacteria operate, created by the air spaces of the aerenchyma (p. 12).

The breakdown of wood litter is slow, but may contribute as much energy and carbon to the ecosystem as leaf litter. If woody debris is not building up, then the rate of breakdown matches the rate of woody litter production. The rate at which breakdown products are made available to other organisms in the ecosystem matches the rate of litter production, even if the intervening processes have taken a long time.

8.1.4.4 *The role of sediment bacteria*

Sediment bacteria are important in facilitating the breakdown of mangrove litter, and are an important element in the flow of carbon through the mangal ecosystem. In the top 2 cm of mangrove sediment there may be up to 3.6×10^{11} bacterial cells/g (dry weight) of sediment, and a productivity of 5.1 g of carbon/m^2 per day, equivalent to an annual production of more than 18 t/ha. These are among the highest values known for any marine sediment (Alongi 1990).

Sediment bacteria are an important food source for meiofaunal species, as well as for deposit-feeding macrofauna. In general, however, meiofauna are not a major element in limiting bacterial populations, or in determining carbon flux rates.

Most bacteria die uneaten, either because they locally exhaust their substrate (local to a bacterium being at a very small scale indeed) or are killed by bacteriophage viruses. When bacteria die, soluble materials are released and add to the pool of dissolved organic carbon available for further bacterial activity. Little of this escapes from the sediment. Dissolved organic carbon is in turn acted on and assimilated by bacteria, and so becomes particulate again. In this way, cycling of carbon may take place entirely within the bacterial community, which constitutes in effect a carbon sink within the mangal. In Malaysian mangroves it has been estimated that, overall, some 1.5 t/ha of carbon accumulate in the sediment each year, about 10% of total production (Robertson *et al.* 1992; Ong 1993).

8.1.4.5 *The fate of organic particles*

Digested by leaf-eating crabs and voided as faeces, or broken down by microbes, much of mangrove production ends up sooner or later as small organic particles (particulate organic matter). Some of this is washed out of the mangal, but the bulk remains. Together with bacterial cells, diatoms,

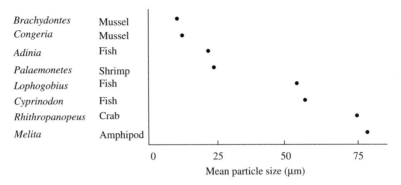

Fig. 8.6 Mean particle size in the stomach or buccal contents of animals in a Florida mangal. Reproduced from Odum, W.E. and Heald, E.J. 1975. The detritus-based food web of an estuarine mangrove community. *Estuarine Research* **1**: 265–286, with permission from Elsevier.

and other matter, it constitutes the food source for fiddler crabs and other deposit feeders, and filter feeders such as oysters.

Fairly obviously, there is a continuum of particle size, from whole leaves (or even fallen trees) to the smallest fleck of organic matter in the mud, or suspended in the water. Figure 8.6 shows that there may also be a continuum of feeders on particles of detritus, each specializing in a different range of particle size.

Of all of the mangrove deposit and filter feeders, the fiddler crabs (*Uca*) are probably the best known. Chapter 6 discussed their population density, sediment-processing rates, and productivity. Clearly fiddler crabs play a significant role in cycling organic matter within the mangal, by ingesting particles and returning faeces to the soil surface. They also contribute to export through their massive production of larvae, as well as by being eaten by commuting predators. Less charismatic deposit and filter feeders are comparatively neglected, but undoubtedly have similarly important roles.

8.1.4.6 Predators

Relatively few macrofaunal predators are permanently resident within the intertidal zone of the mangal: the majority, such as birds, fish, and the larger swimming crabs are highly mobile and commute to and from this zone either tidally, seasonally, or both. Largely for this reason, there is virtually no quantitative information of feeding rates or productivity. Even where such information is available, it is hard to estimate how much of productivity should be attributed to mangrove sources and how much to feeding elsewhere. The rich harvests of crabs, shrimp and fish from waters close to mangroves indicate the general importance of mangrove productivity to these animals (see Chapter 11).

Crocodiles, alligators, and caimans are among the most striking of top predators in mangrove areas. In the Sundarbans of India and Bangladesh, the estuarine crocodile (*Crocodilus porosus*) is an important predator of fish (Chapter 5). It might be expected that these voracious predators would have a distinctly negative effect on fish populations, and hence on human fisheries. In fact, the effect is opposite to that expected: as it happens, crocodiles tend to eat the less commercial fish, some of which in turn feed on smaller fish, and their eggs, of species which are more sought after by humans. Crocodiles, in fact, benefit human fisheries by reducing predators of commercial species. In addition, by their movements they enhance the movement of nutrients in the water, helping to maintain productivity. A similar effect has been noted with caimans (*Caiman crocodylus*) in the Amazon. Wherever caimans disappeared, fishing subsequently declined. In this case nutrients released by caimans in the course of their metabolism, mostly originating from outside the area, were sufficient to have a stimulating effect on primary production (FAO 1994).

These examples give some idea of the complexity of the interactions of predators and prey, and of the possible importance of such interactions, in mangrove ecosystems.

8.1.5 Putting the model together

If the major functional components of the mangrove ecosystem, and the fluxes of carbon between them, have been measured, it is possible to assemble the information into a food-web, or model of the entire ecosystem. Figure 8.7 summarizes the major interactions between the elements of a mangrove ecosystem. Other ecosystems, of course, would show differences in the emphasis on the different components, and it is clear that certain of the components have not been adequately measured.

The example shown is much oversimplified. A number of major components have been omitted, only broadly defined functional groups of organisms are considered, and the model includes only carbon masses and fluxes. A similar diagram could be produced to include the movements and accumulations of nitrogen, phosphorus, or any of the other key nutrients, or, ideally, of all of these together. As it stands, the diagram is perhaps quite complicated enough to give a general idea of the approach.

Why attempt to model ecosystems? In the first place, a quantitative understanding is more satisfying than a qualitative one, and not just as an exercise in accountancy. It informs us, for example, not just that sesarmid crabs as a group eat mangrove leaves, but that that they eat a significant proportion of leaf litter and are therefore crucial to carbon (and energy) flow through the ecosystem. In other mangals, such as those of Florida, this would not be the case, so illuminating comparisons can be made between different areas.

162 THE BIOLOGY OF MANGROVES AND SEAGRASSES

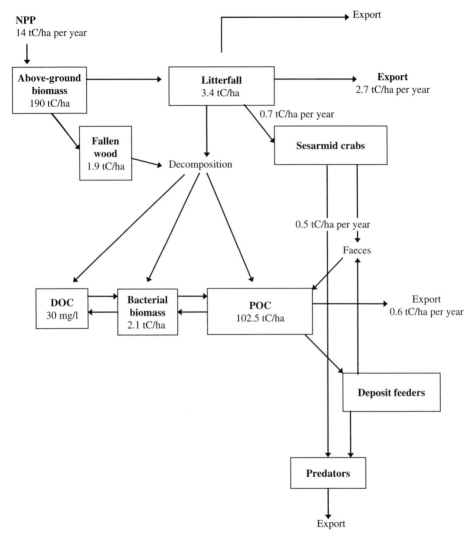

Fig. 8.7 Major carbon fluxes through a mangrove ecosystem. NPP, net primary production; DOC, dissolved organic carbon; POC, particulate organic carbon. Masses are in tonnes of carbon per hectare (tC/ha), fluxes are in tonnes carbon per hectare per year (tC/ha per year). Below-ground biomass and fluxes are omitted. Mainly after Robertson, A.I., Alongi, D.M. and Boto K.G. 1992. Food chains and carbon fluxes. In *Tropical Mangrove Ecosystems*. Coastal and Estuarine Studies, no. 41 (Robertson, A.I. and Alongi, D.M., eds), pp. 293–326. Copyright by the American Geophysical Union.

Secondly, even a simple model can make gaps in knowledge and understanding obvious, and clarify priorities for further research. If a model converges sufficiently closely with reality, it can virtually be used as an experimental system, or at least to suggest fruitful experiments that could be carried out. Finally, a sufficiently detailed model can be used as a powerful planning and conservation tool, in predicting the probable effects of future impacts on a mangrove ecosystem.

One good example is the classical model compiled by Lugo *et al.* (1976), based on the mangroves of southern Florida, and designed to run as a computer simulation. This incorporated not just biomass measurements and carbon fluxes, but solar irradiation, photosynthetic and respiration rates, tidal activity, nutrient availability from various sources and its effect on photosynthesis, and other factors.

Some interesting predictions emerged when this model mangal was allowed to run for a simulated period of many years under various settings of the key environmental variables. If it was started at a simulated early stage of succession, it reached a steady state of biomass in 23 simulated years, at a rather higher biomass (per unit area) than that measured at a real site in the vicinity. Hurricanes occur in the area with a mean frequency of around 20 years, so real mangrove stands will, for the most part, be composed of trees younger than 20 years, with biomass/area levels lower than the maximum potential. The model therefore suggests a possible explanation for the observed values.

Running the model under different conditions of nutrient availability suggested that for characteristic growth rates to be maintained, a steady input of terrestrial nutrients was required, and that nutrient availability rather than salinity was the major limiting factor on mangrove growth. This gave a valuable (and directly testable) insight into mangrove ecosystem function. The information was also directly relevant to management of the Florida mangroves. Freshwater runoff from the land, carrying dissolved nutrients, has been progressively canalized and diverted so as to bypass the mangroves and run directly into the sea. The prediction derived from the simulation model is that this will cause further decline in mangrove growth and productivity.

8.2 Seagrasses

The major features of seagrass ecosystems can similarly be summarized as food-webs or energy-flow models. Seagrass meadows may be tropical or temperate, intertidal or subtidal. They vary more in environmental circumstances than mangroves: nevertheless, some generalizations can usefully be made.

Figure 8.8 shows a generalized tropical seagrass food-web. In its major features, this is similar to a typical mangrove food-web, particularly in relation to the importance of detrital pathways, although the species comprised within each functional group are different.

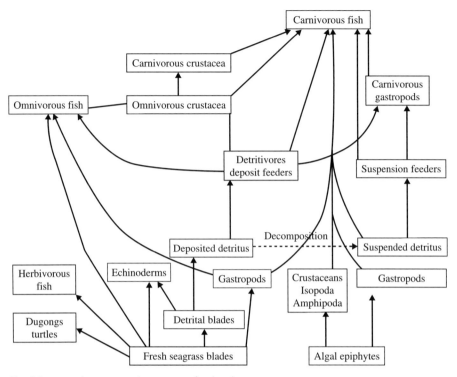

Fig. 8.8 A representative seagrass food-web.

Biomass, energy-flow rates, and other features can in principle be quantified, although in practice this is more difficult to achieve than with mangroves. Fairly obviously, the standing-crop biomass of a seagrass meadow is much lower than that of a mangrove forest. Typical values are less than 5 t of dry weight/ha, although *Amphibolis antarctica* can reach 20 t/ha and a *Thalassia testudinum* meadow, most exceptionally, has been estimated at 42 t/ha (Duarte and Chiscano 1999). The above-ground biomass of a typical tropical *Rhizophora* forest, in contrast, is in the region of 500–550 t/ha (see above). Although seagrass biomass is typically an order of magnitude lower, net primary production is, broadly, in the same range as that of mangroves.

Compared with mangroves, about twice as much of net primary production is directly consumed by herbivores: an average, based on estimates in many different seagrass ecosystems, of 18.6%, compared with 9.1%. In both cases, a substantial proportion of net primary production directly enters detrital decomposition pathways: 50.3% compared with 40.1% (Duarte and Cebrián 1996). These figures are averages, based on many different seagrass ecosystems. At best, such figures are rough generalizations.

Production and biomass vary with latitude (Figure 8.9) and nutrient availability, as well as with such irregular and unpredictable factors as the

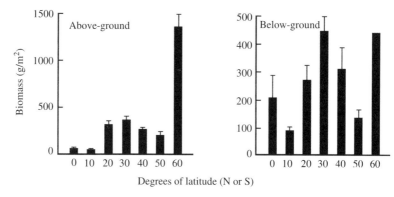

Fig. 8.9 Seagrass biomass (dry weight) in relation to latitude (north or south). Reproduced from Duarte, C.M. and Chiscano, C.L., (1999). Seagrass biomass and production: a reassessment. *Aquatic Botany* **65**, 159–174, with permission from Elsevier.

arrival of a herd of 100 dugongs (Chapter 7). Seagrass plants are small and have a more rapid turnover than mangrove trees, and seagrass meadows respond more rapidly to environmental influences. Seagrass meadows are more dynamic and variable than the rather more sedate mangrove forests.

9 Comparisons and connections

A typical mangrove or seagrass ecosystem is a soft-bottomed intertidal or subtidal community, highly productive, with a diverse fauna. It is dominated by angiosperms adapted to a saline, waterlogged, and anoxic substrate. The emphasis in energy flow through the ecosystem is on detrital breakdown by a combination of detritivores and microbial decomposition pathways, the former complementing and facilitating the latter. In general, control is bottom-up, with primary production regulated by nutrient availability and salinity, secondary producers by primary producers, and so on.

How different are mangrove and seagrass habitats from their immediate surroundings? Are mangroves and seagrass meadows merely areas of substrate with vegetation and consequential concentration of fauna, or distinctive habitats, differing qualitatively as well as quantitatively from their surroundings? Mangroves also show similarities to the predominantly temperate salt-marsh habitats, and comparison of the two sheds some light on the question of why mangroves are exclusively tropical or subtropical.

Even if mangrove and seagrass habitats are distinctively different from their surroundings, they are not isolated from those surroundings. Interchanges take place, of carbon and inorganic nutrients, of eggs and larvae, of commuting invertebrates, fish, mammals, reptiles, and birds. An understanding of these connections places mangroves and seagrasses in their ecological context.

9.1 How distinctive are mangrove and seagrass communities?

Mangrove trees and seagrass plants are distinctive, and define their respective communities. Within these habitats live many other organisms, collectively constituting the community. These are discussed in Chapters 5–7. A key question is whether the species present are so specialized that they cannot live elsewhere, or whether they are drawn from the surrounding pool of species, concentrated in the mangal or seagrass meadow merely because it offers richer resources than those available in the immediate vicinity.

Few species are found only in mangroves or seagrass beds. Among the vertebrates, only a few birds and fewer mammals and reptiles are restricted to mangroves, while fish forage among the trees only at high tide. Many grapsid and ocypodid crabs (p. 102) are regarded as characteristic inhabitants of mangroves, but species actually known to be more or less exclusively found there are surprisingly few: four or five have been found only in mangroves, with a further dozen or so possible candidates. The picture is very similar in seagrasses, with a tiny number of species recorded only from within seagrass beds.

The evidence is not very reliable, partly because biologists working in mangrove and seagrass areas have paid less attention to adjoining mudflats or sand; and, of course, because there is little interest in recording a species' absence from a habitat. One of the few studies focused directly on the faunal specificity is a survey of coastal marine habitats of the entire Saudi coast of the Red Sea. This compared the species composition of seagrass and other littoral and shallow-water habitats, as well as mangroves, and clustered the data from collecting sites by faunal similarity. Mangrove and seagrass habitats showed no greater overall similarity to each other than they did to other soft-bottom habitats.

As far as mangroves (and possibly also seagrasses) are concerned, the Red Sea is not typical. The diversity of mangrove tree species is low (predominantly a single species, *Avicennia marina*), the substrate is usually sand rather than mud with little accumulation of leaf litter, mangrove stands are coastal fringing rather than estuarine, and most are small in extent. These factors are likely to contribute to low diversity of mangrove invertebrates and to make the presence of a specifically mangrove fauna less likely. However, typical or otherwise, here at least there is no evidence of a characteristic mangrove or seagrass fauna. Species are shared with surrounding habitats (Price *et al.* 1987).

The majority of animals in mangroves, and in seagrasses, may therefore be opportunists, facultative rather than obligate inhabitants. Mangrove and seagrass communities are largely assemblages of species attracted by the abundance of nutrients and by the physical habitat offered. If species are, in general, not restricted exclusively to these habitats, individuals are likely to move to and fro between mangroves or seagrasses and their surroundings. If this likely interchange of organisms is taken together with the probable interchange of nutrients and organic matter, it is obvious that these ecosystems cannot be considered in isolation from their surroundings.

9.2 Mangroves and salt marshes

Mangrove habitats are virtually limited to the tropics. In temperate regions, their place is taken by salt-marsh vegetation. Apart from their climatic preferences, how do the two ecosystems compare?

Both mangroves and salt-marsh plants are adapted to live in frequently inundated, waterlogged, saline, and anoxic soils, and both trap and accumulate sediment particles. The most obvious difference between the two is that mangroves are mostly trees, often of considerable height. Salt-marsh vegetation comprises grasses and small shrubs, often only a few centimetres high (although in North America the salt-marsh grass *Spartina* may reach 3 m). There is no obvious reason why tolerance of lower temperatures should impose a height restriction. Salt marshes also tend to be less diverse than mangroves, perhaps due to the decline in species richness with increasing latitude, or to salt-marsh vegetation being less structurally complex.

Because the Atlantic coast of North America has both a large concentration of both salt-marsh habitat (600 000 ha) and biologists, much of our understanding of salt-marsh ecology comes from this part of the world, particularly from the dominant salt-marsh species of North America, the grass *Spartina*. Primary productivity is comparable with that of mangroves, and may be as high as 500 g of carbon/m^2 per year. As with mangroves, there is a tendency for productivity to decline with increasing latitude. Productivity also declines with increasing salinity, as it tends to with mangroves. There is some doubt as to whether salt-marsh growth and productivity are normally limited by salinity levels, or by limitation of nutrients, particularly of nitrogen (Mann 1982).

The fate of salt-marsh production diverges somewhat from that of mangroves. Direct herbivory generally accounts for a greater proportion, more is retained within the ecosystem, and less is exported (Table 9.1). Leaf-shredders do exist, but are less important than, for instance, the sesarmid crabs of the Indo-Pacific mangal. Deposit-feeding fiddler crabs (*Uca* spp.), on the other hand, may be relatively more important. Certainly

Table 9.1 Comparison of the fate of primary production in mangroves and salt marshes. The figures are in percentages, but because each is an average of a number of independent estimates, they do not necessarily total 100%. From Duarte and Cebrián (1996).

	Fate of primary production (%)	
	Mangrove	Salt marsh
Eaten by herbivores	9.1	31.3
Exported	29.5	18.6
Decomposed within system	40.1	51.2
Stored in sediments	10.4	16.7

their abundance is higher: concentrations of up to 200 crabs/m^2 have been recorded (Macintosh 1982). The burrowing activity of salt-marsh *Uca* has a major effect on salt-marsh primary production. Reduction of *Uca* population density in one case reduced above-ground production by 47% and increased root-mat density by 35% (Bertness 1985).

Finally, salt marshes may be economically as well as ecologically important habitats, for similar reasons to mangroves. They play a major role in protecting against coastal erosion, and sustain coastal shrimp and other fisheries.

9.3 Interactions

To what extent do mangroves and seagrasses exchange materials with their surroundings? With mangroves this varies considerably, depending on the environmental setting. At one extreme, the rare mangroves which grow in complete isolation from the sea (p. 52) are unlikely to export much of their productivity to surrounding habitats. The most significant means of export would be by animals feeding within the mangal and subsequently moving elsewhere. At the opposite extreme, a mangrove flushed by a large river, or by frequent tidal inundation and active currents, is likely to lose a much greater proportion of its litter production. Rivers and tides can also bring in nutrients and particulate matter. Soluble nutrients (particularly nitrates and phosphates) can then be retained and used by the mangrove trees, and organic particles trapped in the mangal (p. 64), so that imports as well as exports may be relatively high.

Interaction between habitats can also result from the movements of animals, through the dispersal into the open sea of planktonic larvae and their subsequent return, or by movement between habitats seasonally, or at different stages of the life cycle. These can represent considerable transfers of material. The amount of inorganic nitrogen and phosphate deposited by 220 000 roosting fruit bats (p. 96), the annual removal of carbon and nitrogen represented by a flourishing fishery, or the amount of primary production harvested by a roving herd of dugongs, are not negligible. These could be estimated with reasonable accuracy, but it is more difficult to assess the influence of transient visitors such as fish or shrimp commuters, or of larval dispersal and return.

The following sections discuss some of the most significant interactions between habitats: the movement from mangroves of organic matter as dissolved and particulate carbon, the dispersal and return of larvae, and the movement of juvenile and adult fish between mangroves, seagrasses, and coral reefs. Finally, a practically important question of the movement of material from seagrasses and mangroves is the extent to which commercial fisheries depend on these habitats.

9.4 Outwelling

Early studies of salt marshes indicated that they were highly productive, and that much of this production was probably exported, where it sustained secondary consumers and food chains in nearshore waters contributing significantly to commercial fisheries. The proposition that productive coastal ecosystems were net exporters of organic matter to adjacent waters was soon extended to mangroves. Outwelling was seen as an important feature of mangroves, and a strong argument for their conservation.

The outwelling hypothesis rests initially on a fairly simplistic balancing of the ecological books. Primary production can be measured, partially in terms of litterfall, or more fully by the methods discussed in the last chapter. Consumption by herbivores, and decomposition within the mangal, can also be measured, although probably with less reliability. If production is greater than consumption and *in situ* decomposition, then the difference can be assumed to represent export from the mangal. The estimate can be made more realistic if net accumulation of matter within the mangal is considered. Imprecision in the original measurements is compounded in the final figure but, failing any fuller data, this mass balance approach can give a reasonable provisional indication of the movement of organic matter out of the mangroves.

This approach inferred export indirectly from apparent shortfalls between estimates of production and of fate. Export can also be estimated directly, by combining information on the movements of water through the mangal with measurements of its content of particulate and dissolved organic carbon. Mangrove habitats are often topologically complex, and water movements vary with currents, tides, and meteorological conditions. To give a complete picture, measurements have to be carried out over a prolonged period of time, to include seasonal and other variation. A brief rainy season may flush out leaf litter or inorganic substances which have been steadily accumulating, resulting in a pulse of exported nutrients to the surroundings which would be missed in a short-term study. Merely collecting the raw data, before assembling it into a full model of carbon movement, is therefore a task of Herculean proportions. Not surprisingly, few such comprehensive studies have been achieved.

As usual, one of the most thorough analyses comes from the mangroves of Hinchinbrook Island, Queensland, in this case a study of the tidal mangrove forests on Coral Creek, Missionary Bay. A proportion of the litterfall was sequestered by crabs: after allowing for this, it was estimated that about 7.5 kg of carbon/ha per day was exported as recognizable leaf litter, and a further 1.6 kg of carbon/ha per day in the form of particulate organic carbon. Together, these components comprised around 30% of leaf-litter production. Dissolved organic carbon was also measured, but exchange of dissolved organic carbon was trivial in comparison, amounting to a net import of only 0.2 kg of carbon/ha per day. Labile forms of dissolved organic carbon appear to be efficiently recycled within the forest, rather than exported (Boto *et al.* 1991; Robertson *et al.* 1992).

A contrast to the productive mangroves of Australia is shown by high intertidal forests in Florida, where the total export of particulate carbon was only 0.4 kg of carbon/ha per day. Here dissolved organic carbon was of relatively more significance, amounting to an export of 1.0 kg of carbon/ha per day, more than 70% of the total carbon exported (Twilley 1985).

As already discussed, the fate of primary production varies greatly, depending in particular on hydrological conditions. Riverine mangroves may acquire an input of particulate and dissolved organic carbon with the river water, as well as contributing their own productivity to the outflow, and a tidally dominated mangal is likely to export more of its production than a high shore, or even inland one (see Figure 8.4). Except in a very few cases, there is simply insufficient information to establish the scale and significance of organic carbon outwelling; the only valid generalization is that generalization is impossible.

The position with respect to other materials, particularly nitrogen and phosphorus, is even less clear. Again, Missionary Bay provides the most thoroughly investigated example. The principal sources of nitrogen are the products of nitrogen fixation within the mangal, and dissolved organic nitrogen, nitrate, nitrite, and ammonium brought in by tidal flow. Rapid recycling of nitrogen takes place within the mangal (see Chapter 2), and of course large amounts of nitrogen are static there in the form of standing crop biomass—perhaps 6000 t of nitrogen in above-ground biomass alone. Both soluble and particulate nitrogen are exported, with a considerable net import of soluble forms of nitrogen, and a net export of particulate nitrogen (Table 9.2). As far as available information goes (which is not very far), this seems fairly typical of mangroves. The phosphorus budget is less clear, but there appears to be a significant net annual import, corresponding to nearly a quarter of what is required for the known rates of primary production (Boto *et al.* 1991; Alongi *et al.* 1992).

Table 9.2 Nitrogen budget from Missionary Bay mangrove forest, Queensland. Negative figures in the balance column indicate net export. Data modified from Alongi *et al.* (1992).

Process	Nitrogen budget (kg of nitrogen/year)		
	Input	Output	Balance
Nitrogen fixation	36 831		36 831
Denitrification		2824	−2824
Tidal: dissolved	168 598	116 108	52 490
Tidal: particulate		76 321	−76 321
Totals	205 487	195 253	10 234

9.5 The fate of mangrove exports

Given the extent of carbon export from mangrove ecosystems, marked effects on neighbouring ecosystems would be expected. What is the fate of the outwelled carbon, both particulate and dissolved, and what are its effects?

Physically identifiable mangrove detritus does not generally go far. In one study of mangroves in southern Florida, for instance, the percentage of mangrove material in plant litter fell off from 80% to about 10% within less than 100 m of the edge of the mangrove forest (Figure 9.1; Fleming *et al.* 1990). Although dispersal will depend on local conditions of river flow, tides, and currents, leaf litter probably has only very local effects.

Much of the exported organic material is in the form of small organic particles, or dissolved matter, which are not so readily identifiable. Dissolved organic matter from mangroves can be identified by the analysis of lignin-derived substances, and naturally fluorescing compounds which could have come only from vascular plants. On this basis, about 10% of the dissolved organic matter detected in the coastal waters of the Bahamas appeared to have come from mangroves at least 1 km away. Correlation between the concentration of mangrove-derived dissolved organic matter and bacterial production suggested that this contributed significantly to bacterial production in the plankton (Moran *et al.* 1991).

Measurement of the ratio of stable isotopes of carbon has been widely used as a technique for tracking carbon through food chains (see p. 105). The ratio, relative to a standard, is expressed as a $\delta^{13}C$ value, in parts per thousand. Mangrove leaves have $\delta^{13}C$ values in the range -24 to $-30‰$, while other

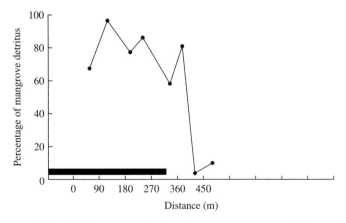

Fig. 9.1 Percentage of identifiable mangrove detritus along a transect outwards from a riverine mangrove forest into a seagrass bed in southern Florida. The horizontal bar indicates the extent of the mangrove forest. Reproduced with permission from Fleming, M., Lin, G., and Sternberg, L.da S.L. 1990. Influence of mangrove detritus in an estuarine ecosystem. *Bulletin of Marine Science* **47**: 663–669.

marine plants have a higher (less negative) $\delta^{13}C$ values. The isotope composition of a consumer reflects that of its diet, with only a slight shift due to selective handling of the two isotopes during respiration.

Figure 9.2 shows the range of $\delta^{13}C$ values in organic matter from sediments, and in the tissues of animals collected from mangroves and their associated mudflats in western Malaysia, from coastal inlets within 2 km of the mangroves, and in offshore waters between 2 and 18 km away. Even animals

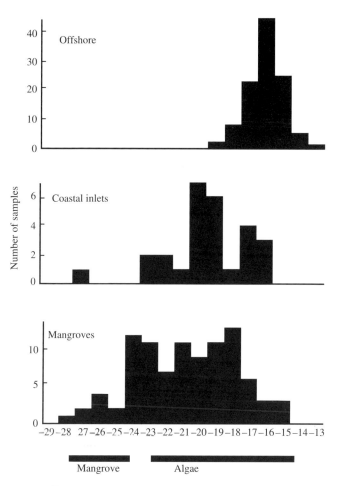

Fig. 9.2 Distribution of $\delta^{13}C$ values (‰) in the tissues of animals caught within a Malaysian mangrove swamp, in coastal inlets among the mangroves, and offshore. The horizontal bars below indicate the range of $\delta^{13}C$ values from mangrove leaves and algal material. Values of sediment material were -24.8 ± 0.9‰ (within mangrove forest), -26.2‰ (mudflat), and -24.8‰ (5–6 km offshore). From data in Rodelli, M.R., Gearing, J.N., Gearing, P.J. et al. 1984. Stable isotope ratio as a tracer of mangrove carbon in Malaysian ecosystems. *Oecologia* **61**: 326–333, with kind permission from Springer Science and Business Media.

collected from within the mangal itself have $\delta^{13}C$ values, on the whole, which differ from the range found in mangrove material. This shows that their primary carbon source is not mangal material. Any influence of mangrove carbon wanes still further in nearshore waters, where the principal carbon source, on this evidence, seems likely to be phytoplankton, whereas offshore animals seem on this basis to have no dependence whatever on mangrove-derived carbon (Rodelli et al. 1984).

The $\delta^{13}C$ values of carbon in nearshore and offshore sediment samples is closer to that of mangrove material, and it appears that a large proportion of mangrove carbon is initially deposited in sediments. Returning to the studies on Missionary Bay, isotope studies were used to estimate the fate and disposition of mangrove carbon after it was exported from the mangal. Sites 1–2 km from the mangal accumulated at the rate of around 440 kg of carbon/ha per year, and even at sites 10 km from shore the accumulation rate was 10–40 kg of carbon/ha per year. This accumulation of material appeared to have little stimulating effect on bacterial or meiofaunal activity. The mangrove material appears to be either refractory, possibly because of high residual tannin levels, or simply of little nutritional value until colonized by bacteria (Robertson et al. 1992).

Interpretation of stable isotope ratios is not always straightforward. The general inference that consumers collected within the mangal do not depend directly on mangrove carbon depends on which species are collected. Values of $\delta^{13}C$ higher than those of actual mangrove leaf material, but lower than those of alternative primary producers, suggests that the species sampled utilized more than one carbon source in their diet, rather than that they do not use mangrove-derived carbon. Finally, tissue $\delta^{13}C$ values depend on the number of intervening stages in the food chain as well as on the source of primary production.

With reservations, however, it can be deduced that much of the carbon exported from mangroves (at least from the few sites studied to date) settles initially in sediments within only a few kilometres. Here it is processed slowly by bacterial action. Possibly successive cycling through bacterial metabolism erases the $\delta^{13}C$ signature characteristic of mangrove material. Eventually much of the carbon is released into the water column as dissolved organic carbon and mixed with carbon from other sources, and in this form it may re-enter planktonic and macrofaunal foodchains, either in the same location of dispersed over a much wider area.

9.6 Larval dispersal and return

Species may alternate between mangrove and other habitats at different stages of their life cycle. Many coastal species disperse their offspring as planktonic larvae, and have juveniles which settle in and exploit different habitats from adults, as nursery areas.

The planktonic zoeae larvae of mangrove crabs have already been discussed (p. 98). Vast numbers of these are produced, and released into mangrove creeks or channels, where they are carried away to pass their larval life in open waters nearby. After a series of moults, a zoea metamorphoses into a settling form, the megalopa, then into a miniature crab. Similarly, sessile animals such as barnacles or bivalve molluscs also have planktonic larval stages.

A single female fiddler crab (*Uca*) releases tens or even hundreds of thousands of larvae each year, representing quite a significant export of organic matter from the mangal to nearshore waters. Of these larvae, probably only at most a few hundred survive and return as megalopae. The fact that even this number return, however, is quite remarkable. If the currents carry zoeae passively out to sea, how do megalopae later contrive to return?

The answer is that planktonic larvae are not simply passive organic particles, to be swept wherever current and tide may take them. Larvae often remain not far from, and return to, the area from which they originated. To understand how they achieve this, it is necessary to look more closely at the pattern of water flow in a coastal system such as an estuary.

In an estuary there is a net outward flow of fresh water. Often, this moves as a surface layer over an underlying wedge of denser sea water. At the interface between the fresh and salt water layers, shear causes a degree of mixing. There are two consequences of this: the salinity of the surface water increases as it moves seawards, and to compensate for the entrained salt water, a slow upstream current of deeper, salt water is generated. Where there is a strong tidal flow, more extensive mixing of the two layers takes place, more salt water is entrained, and the upstream current is greater.

This combination of downstream and upstream water movements can result in the retention of larvae and their return to a site not far away from their point of origin. Early larvae maintain a position near the surface of the water, and are transported to the mouth of an estuary, or even out to waters of the continental shelf. After a time, they sink, and at deeper levels return upstream by riding the incoming current. Although this transport mechanism has not been demonstrated for any mangrove species, it has been shown to occur with a number of salt-marsh species, including fiddler crabs (Epifanio 1988).

9.7 Mangroves, seagrasses, and coral reefs

Seagrass beds often grow in close proximity to mangroves and coral reefs, and the ecosystems are often closely linked through fluxes of carbon and other materials.

Links between mangroves and seagrasses have again been elucidated by studies using stable isotopes. Fortunately, seagrasses show $\delta^{13}C$ values much higher (less negative) than those of mangroves, making it possible to identify an influence of mangrove carbon within a seagrass bed. In southern Florida, fronds of the seagrass *Thalassia* growing close to mangroves had more negative (that is, more mangrove-like) $\delta^{13}C$ values than those collected far from mangroves. The same was true of carbon in calcium carbonate of mollusc shells collected from sites near to, and far from, the mangroves (Table 9.3). This strongly suggests a short-range influence of mangrove carbon, probably through the assimilation by the seagrasses (and subsequently molluscs) of carbon dioxide released by the processing of mangrove material (Lin et al. 1991).

Similar results from a mangrove forest and adjacent seagrass beds in Kenya suggested a similar conclusion. Here there was a gradient in $\delta^{13}C$ values from the mangroves through the beds of seagrass (*Thalassodendron*), both in the sediments and in seagrass leaves. The most probable interpretation is that particulate organic matter from the mangroves is distributed in the sediments in amounts which decrease rapidly with distance from the source, and that microbial respiration in the sediment then makes CO_2 from the particulate organic matter available for seagrass photosynthesis. During flood tides, particulate organic matter appears to flow back from the seagrass beds into the mangal. The two ecosystems are thus tightly coupled by reciprocal movements of carbon (Hemminga et al. 1994).

Connections between mangroves and coral reefs are more elusive, but may be important. Corals are very vulnerable to sedimentation and to eutrophication, which results from nutrient enrichment. In some situations, therefore, mangroves may protect reefs by trapping river sediments, and sequestering nutrients. Mangroves, seagrasses, and reefs all tend to export organic matter, both dissolved and particulate (dissolved and particulate organic matter). There appears to be a net movement of dissolved organic matter in the direction of mangroves→seagrasses→coral reefs. The situation with particulate organic matter is more complex, and neither dissolved

Table 9.3 $\delta^{13}C$ values for seagrass leaves and mollusc carbonate collected close to, and far from, mangrove stands. The $\delta^{13}C$ value for mangrove carbon normally falls in the range -24 to -30‰. From Lin et al. (1991).

	$\delta^{13}C$ value (‰; ±S.E.)	
	Near	Far
Seagrass leaves	-12.8 ± 1.1	-8.3 ± 0.9
Mollusc shell	-2.3 ± 0.6	0.6 ± 0.3

organic matter nor particulate organic matter movement has been adequately quantified.

9.8 Commuting and other movement

Passive transport of dissolved and particulate organic matter is not the only connection between habitats. Considerable amounts of organic matter move to and fro actively, particularly in the form of fish and crustaceans, and although these movements are difficult to assess quantitatively in terms of carbon flux, they are nonetheless important.

A habitat can be exploited by a species which spends only part of its time in that habitat. Individuals might commute in periodically to feed, or to use the mangrove or seagrass environment as a refuge from predators, for example at certain stages of the tide, or as a key habitat for only certain stages of the life cycle, perhaps as a nursery area for larvae or juveniles.

Fish are abundant in mangrove creeks (Chapter 6). The biomass of fish in creeks and channels within the mangal may be several times as great as that of similar channels in a mudflat area. At high tide, many fish invade the flooded mangrove forest itself, and analysis of stomach contents shows clearly that they are actively foraging for food there (Table 9.4) As many of these commuting fish range over adjacent habitats and inshore waters, they represent an important functional link between mangrove and other habitats.

Mangroves and seagrass beds may be particularly closely linked. Given the similarities in the two ecosystems, it is not surprising that they share some of their fauna, and that faunal movements between mangroves and seagrasses represent a significant functional link, supplementing the coupling of the two ecosystems by movements of particulate organic matter. In particular, many species of fish seem to move freely between seagrass beds and mangrove forests, presumably feeding in both. In the Gazi Bay area of Kenya, $\delta^{13}C$ estimates indicated that the fish species sampled fell into three general categories. One group occurred entirely within the seagrasses, and had

Table 9.4 Composition of the gut contents, by volume, of four species of fish caught within a Malaysian mangrove forest, to show dependence on plant material and on mangrove animals. Data from Sasekumar et al. (1984).

Fish species	Composition of the gut contents, % by volume						
	Mangrove	Bivalve	Gastropod	Prawn	Crab	Fish	Other
Tachyurus	1	3		3	71		21
Arius	1		2	27	46	17	7
Pomadasys			2	16	81		1
Liza	38			23			39

tissue $\delta^{13}C$ values that fell entirely within the range for seagrass carbon (-10.07 to $-19.82‰$). A second group were found in mangrove creeks, but again had tissue $\delta^{13}C$ values indicating seagrass dependence. The third group of species had $\delta^{13}C$ values intermediate between those of mangrove and seagrass, and presumably depended on both habitats (Marguillier et al. 1997).

Both mangroves and seagrasses serve as nursery areas for juveniles of many coral-reef fish, which often shift between these habitats in complex ways. In Belize, for example, juvenile bluestriped grunt (*Haemulon sciurus*) settle initially in seagrass beds, later shifting to mangroves before their adult life on coral reefs. They are not totally dependent on mangroves, however, and in regions in which mangroves are scarce juveniles graduate directly to reefs. In contrast, juvenile rainbow parrotfish, *Scarus guacamaia* (Scaridae), are dependent on mangroves (Mumby et al. 2004). Other fish show even more complex and variable patterns of use of mangrove and seagrass habitats, sometimes using them for shelter rather than for feeding, sometimes for both. Sometimes different individuals of the same species adopt different strategies, either sheltering in mangroves and foraging in both mangroves and seagrasses, or using seagrasses as both refuges and feeding areas (Nagelkerken and van der Velde 2004). Overall, it appears that mangroves are more significant nursery areas in the Caribbean, as are seagrass beds in the Indo-Pacific (Dorenbosch et al. 2005).

9.9 Mangroves, seagrasses, and fisheries

9.9.1 Shrimps

The relationship of mangroves and seagrass beds to their surroundings is of more than academic interest when commercial fisheries are considered. To what extent do they actually depend, directly or indirectly, on mangrove or seagrass primary production? Even species caught in offshore fisheries may rely on outwelled material, or may depend on mangroves or seagrasses for breeding, or as a nursery habitat at certain stages of their life cycles.

This applies equally to fish and to shrimps (or prawns: the terms are used interchangeably), although the greatest efforts have been to establish links between mangroves and shrimps. The shrimp–mangrove relationship also crops up in the context of aquaculture, where in parts of the world large areas of mangrove have been destroyed to construct aquaculture ponds for mass rearing of shrimps. If shrimps depend on mangroves, potential drawbacks to this procedure are predictable. Some of the issues raised by aquaculture are discussed further in Chapter 11.

Most of the commercially fished or cultivated shrimps belong to the family Penaeidae. The geographical distribution of penaeid species fits the geographical distribution of mangroves 'like a hand in a glove' (Chong 1995).

The northerly and southerly limits of nearly all penaeids is marked more or less by the 15°C winter isotherm, as is the distribution of mangroves by the 20°C winter isotherm (see Chapter 1). Where there are mangroves there are penaeids. The reverse is not necessarily true: to the north and south of the mangrove zone, penaeids may occur in significant numbers in the absence of mangroves, examples being areas such as Florida, California, the Mediterranean, and south-west Australia. Here, however, the place of mangroves may be taken by salt marshes or extensive seagrass beds, so penaeid populations may still be dependent on vascular plants as a primary productivity base.

Areas with particularly well-developed mangroves seem to be associated with particularly rich stocks of shrimp. The islands and channels of the Klang Strait, Malaysia, have luxuriant mangrove forests, and the adjacent coastal waters provide an annual harvest of 10 500 t of shrimp (Chong *et al.* 1996). This, and similar observations, have given rise to several attempts to quantify the relationship between mangroves and shrimps by correlating catches with mangrove area. In some cases, statistically significant linear relationships have been established. In western Malaysia, for instance, the annual shrimp catch was related to mangrove area by the formula

$$C = 0.6368 + 0.5682A$$

where C is the annual catch (t×10^3) and A is the mangrove area (ha×10^3). Similar relationships have been derived in other areas.

Some studies have concluded that the dependence of shrimps on mangroves is weaker and more local than is sometimes supposed, and applies only to certain penaeid species; particularly to those that are known to use mangroves as nursery areas (*Penaeus merguiensis*) rather than to those that do not (*Penaeus semisulcatus*). Attempts to correlate mangrove vegetation with fisheries statistics generally do not distinguish between species (Lee 2004).

Shrimp catches tend to decrease at higher latitudes, and this observation has been incorporated into a general formula relating the maximum sustainable yield (MSY) of a shrimp fishery to both mangrove area (A) and latitude ($L°$):

$$\log_{10}(\text{MSY}) = 2.41 + 0.4875 \log_{10}A - 0.0212L$$

MSY is the estimated maximum catch that can be harvested without jeopardizing the population being exploited (Chong 1995).

What do such correlations mean? The inference that has generally been drawn is that there is a causal connection; that shrimps are so dependent on mangroves that the size of a shrimp population (or annual catch statistics as a substitute for a population estimate) is determined by the area of contributing mangroves. Correlation does not necessarily mean causation, however—as demonstrated by the well-known case of the excellent correlation between the stork population of western Germany and the human birth rate there. Do shrimps actually depend on mangroves, or is the close mathematical relationship a consequence of both being determined by a third variable? Extensive mangroves and sizeable shrimp populations also correlate with large river estuaries, so both could independently be affected by the availability of fresh water, or of silt or nutrient flowing into the sea. What are the connections between shrimps and mangroves?

A causal connection between mangrove area and offshore shrimp catches could result from shrimps being dependent on outwelled organic matter, or on the shrimps using the mangal at a certain stage of their life cycle, for example as a nursery area. As discussed above, organic matter outwelled from mangroves does not seem to travel very far, and shrimp are caught so far offshore that it seems unlikely that they could benefit directly from it.

Studies using stable isotopes confirm this. Figure 9.3 shows the distribution of $\delta^{13}C$ values in penaeid shrimps in Malaysia. These were caught both offshore, and in creeks and inlets within the mangal. It appears that shrimp within the mangal depend partly, but not entirely, on mangrove carbon. Those collected offshore, in contrast, show no sign of having assimilated carbon directly from mangrove material (Rodelli *et al.* 1984).

An alternative interpretation is that offshore shrimps depend on mangroves at early stage of their lives and later move out to offshore waters. As they feed and grow, the mangrove $\delta^{13}C$ signature becomes swamped by later assimilation of carbon from other primary producer sources.

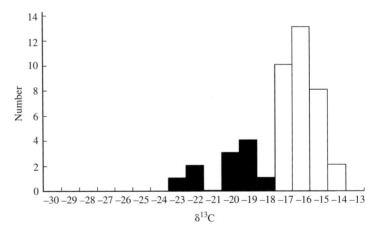

Fig. 9.3 Distribution of $\delta^{13}C$ values in penaeid shrimps caught in coastal inlets in a Malaysian mangrove forest (black bars), compared with individuals caught 2–18 km offshore (white bars). The range of $\delta^{13}C$ values in mangrove leaf material was −28.5 to −24.5‰. From data in Rodelli, M.R., Gearing, J.N., Gearing, P.J. *et al.* 1984. Stable isotope ratio as a tracer of mangrove carbon in Malaysian ecosystems. *Oecologia* **61**: 326–333, with kind permission from Springer Science and Business Media.

In many crustacea, the female carries eggs attached to her abdomen until they hatch as quite advanced larvae, or as juveniles. Penaeids are unusual in this respect. Females may produce as many as 500 000 eggs, and release them directly into the water. Hatching takes place within a few hours, producing larvae called nauplii. The nauplius larva moults several times, before metamorphosing into a different form of larva known as a protozoea. These are filter feeders, dependent on phytoplankton. Neither nauplii nor protozoeae look remotely like adult shrimps. After three or so protozoeal stages, larvae metamorphose again, this time into a more shrimp-like larva, the mysis. After a further three moults (typically, making 11 stages in all) the mysis changes again into a postlarva, very similar in body form to the adult shrimp.

Although all penaeids show similar patterns of development, some undergo this elaborate sequence of larval development within an estuary, while in others the life cycle is entirely pelagic. Most species, however, move between offshore and inshore waters at different ages. In many of the commercially important species, spawning takes place at depths of 10–80 m in offshore waters. The larval stages, as always, are planktonic, but postlarvae and juveniles are abundant in mangroves. In some cases, they actively migrate shorewards. In others, it seems they are carried more or less passively and end up in mangroves as the consequence of current and tidal movements. In the Klang Straits of Malaysia, for instance, it has been demonstrated that strong tidal currents favour the shoreward movement of penaeid larvae, which become trapped in the mangal by physical movement of water along mangrove-fringed channels. Around 65 billion larvae are retained in the Klang mangal by this mechanism (Chong *et al.* 1996).

Virtually all shrimps found in mangrove areas are juveniles, so mangroves clearly merit the term nursery areas. What do young shrimp gain from their residence in the mangal?

The first, and obvious, benefit is nutritional. While in the mangal, juvenile shrimps both ingest and assimilate mangrove detritus: this is confirmed by analysis of gut contents, and by stable-isotope studies (Figure 9.3). The gut contents of Malaysian penaeids may comprise 40% mangrove material, and stable isotope analysis suggests that around 65% of their carbon originated from mangrove detritus (Rodelli *et al.* 1984). Apart from direct intake of mangrove detritus, mangrove creeks provide a rich habitat within which are many other suitable forms of food for omnivorous shrimps.

Mangroves (and seagrasses) also provide shelter and protection. At high tide, juvenile shrimps enter the flooded mangrove forest and feed there. When the tide ebbs, they resist the flow until the water level (and available forest habitat) are quite low, then enter the creeks and congregate in turbid shallow waters close to the water's edge at low tide. When the tide rises, the shrimps again move in among the inundated trees. Although this behaviour is related to foraging in the flooded forest, it also reduces the risks of predation. Many fish feed on juvenile shrimp (Table 9.4), and it has been shown experimentally that the structural complexity of dense pneumatophores greatly reduces the efficiency rate of predatory fish, and the mortality rate of shrimps (Primavera 1997a).

9.9.2 Fish

Even a few hours with a cast-net fisherman in the mangrove creeks of Malaysia leaves a lasting impression of the richness and diversity of the fish population. Typical tropical mangroves in the Indo-Pacific region may have close to 200 species. Temperate mangroves, and those in the Atlantic, tend to be rather less species-rich, but even here more than 100 species have been recorded from a single location. Seagrass beds, too, may be rich in species of fish.

Analysis of fish stomach contents confirms the dependence of fish on mangrove invertebrates, and on smaller fish, for food. The majority of mangrove fish are juveniles of species which spawn elsewhere, so that fish clearly use mangroves as nursery areas in much the same way as shrimps do. Finally, the structural complexity of the mangal affords shelter from predators. The abundance and diversity of fish is closely related to the amount of mangrove debris present (Robertson and Blaber 1992). In Sri Lanka, this is exploited by local fishermen, who construct artificial thickets of mangrove branches within mangrove lagoons. After a few weeks these attract and concentrate large numbers of fish, which can then be conveniently trapped.

The dependence of commercially caught fish species on mangroves and seagrasses varies with circumstances, with the species of fish in question, and with geographical position. For some species—such as the barramundi *Lates calcarifer* in north-eastern Australia—the link is strong, but for other coastal species evidence of dependence on mangroves is tenuous or negative (Manson *et al.* 2005).

There is a growing awareness that the relationships between fisheries species and mangrove and seagrass beds is complex, and can only be fully understood by taking account, not just of the area of mangrove and seagrass habitats, but of the proximity of these to each other; by taking a landscape approach, and understanding the complexity of a heterogeneous environment at a large scale (Pittman *et al.* 2004).

10 Biodiversity and biogeography

This chapter considers the intersecting questions of biodiversity and biogeography. What determines the diversity and distribution of mangrove and seagrass species throughout the world?

The overall distribution of a group of species such as mangroves or seagrasses comprises the separate ranges of many individual species. These cannot be explained purely by physical or climatic limitations: similar climatic regions in different oceans are occupied by different species. Biogeography raises questions of the origin and spreading of species: when and where did mangroves and seagrasses originate, and what factors led to their present distribution? Much of the present distribution pattern can be explained only by taking account of evolutionary events: the past is the key to the present.

Biodiversity is an elusive concept. Before considering how and why it varies, and the significance of that variation, the concept must be explored and, if possible, clarified. Important questions of scale are involved: biodiversity can be considered at many scales, from molecular to global.

The actual number of species (and of associated organisms) differs, not just geographically, but between sites within a geographical region. To what extent does biodiversity depend on biogeographical or evolutionary factors, and to what extent on local physical or biotic conditions?

And does biodiversity actually matter? Theoretical considerations suggest that biodiversity might have important influences on ecosystem function and stability. What is the functional significance of biodiversity?

10.1 What, if anything, is biodiversity?

Biodiversity has become common currency with the media and with politicians. In the process, a complex concept has come to be used (and misused) in an oversimplified way. It has been caustically defined as 'biological diversity—with the logical part taken out'. What *is* biodiversity?

In its simplest sense, it is often taken to mean simply the number of species present, or species richness. Even this sense needs qualification. The area in question must be defined: all other things being equal, the greater the area being considered, the greater the number of species that will be encountered within it. Species–area relationships have been shown for many groups of organisms, over a wide range of spatial scales. The precise relationship between species and area—typically a simple exponential one—is much more informative than simple species catalogues, but in most cases the information has not been assembled in an appropriate way, and species listings must suffice.

Even simple species counts have hidden complications. Relative abundance of species in some sense contributes to diversity. Two patches of habitat might each contain 100 individuals. In one, the 100 individuals comprise 10 of each of 10 different species. In the other, 10 species are also present: in this case one species dominates, accounting for 91 of the 100 individuals, with the other species each represented by a single individual. The second habitat is surely less diverse than the first. A further complication arises if taxonomic levels are taken into account. A habitat containing 100 species of insect, all of which are beetles, is less diverse than one with 100 species of widely different affinities. Finally, of course, intraspecific diversity, particularly genetic variation within a species, is of enormous evolutionary importance.

Species diversity at a specific point is sometimes referred to as alpha diversity. Introducing a spatial dimension and recording the accumulation of species with distance as we move away from a point has been termed beta diversity. Species–area curves are one method of measuring beta diversity. Gamma diversity refers to the number of species in a whole region. These terms were in vogue for a time, and are still encountered occasionally, but are generally now regarded as unnecessary jargon (Rosenzweig 1995).

As far as mangrove and seagrass biodiversity is concerned, pragmatism dictates the basic estimation of species number as the most useful approach.

10.2 Mangroves

10.2.1 Regional diversity

The geographical distribution of mangrove species, and of mangrove habitats, is limited by temperature. More specifically, the 20°C winter sea-temperature isotherm, with few exceptions, circumscribes the range of mangroves throughout the world (Figure 1.1). Possible reasons for this imitation have been discussed elsewhere in the book (p. 38).

If the latitudinal range of mangroves is curtailed by temperature, the longitudinal distribution of species is affected primarily by physical barriers. The tropical coastlines of the world are separated by the great land masses, and by open oceans, into two major regions and a number of smaller ones. From the point of view of mangroves, the primary separation is between the Indo-West Pacific (IWP) and the Atlantic–Caribbean–East Pacific (ACEP) regions. Africa, and the cold waters round the southern tip of Africa, effectively prevent contact between western Indian Ocean and eastern Atlantic tropical waters.

The barrier between the western and eastern Pacific is less obvious. It consists, quite simply, of the largely empty space of the central Pacific, which forms a barrier to mangrove dispersal. The number of species of mangrove declines eastwards across the Pacific, the easternmost limit being Samoa (170°W). Beyond this, no mangroves occur naturally until the west coast of the Americas. If artificially introduced to intermediate islands (for example to the Society Islands or Hawaii) they thrive, showing that their absence is due to inaccessibility, not to a lack of suitable habitats. A single species, *Rhizophora samoensis*, appears to have crossed the Pacific in the reverse direction, spreading from the Pacific coast of South America as far as New Caledonia (Figure 10.1). It is virtually indistinguishable from

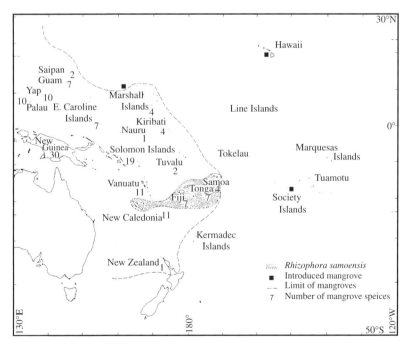

Fig. 10.1   The distribution of mangrove species (and recognized hybrids) in the western Pacific. Reproduced with permission from Woodroffe, C.D. 1987. Pacific island mangroves: distribution and environmental settings. *Pacific Science* **41**: 166–185. ©University of Hawaii Press.

the American species *Rhizophora mangle*, presumably a recent ancestor (Tomlinson 1986; Woodroffe 1987).

The ACEP region falls into three subregions: the eastern Pacific; the Caribbean and western Atlantic separated from it by the American continent and the Isthmus of Panama; and the Atlantic coast of West Africa, separated from the western Atlantic by open ocean. Similarly, the IWP region is biogeographically separated into East Africa (and the Red Sea); an Indo-Malesian subregion comprising the Indian subcontinent and west and south-east Asia; and Australasia, consisting of Australia, New Zealand, New Guinea, and the islands of the western Pacific. These six subregions are shown in Figure 1.1.

The IWP and ACEP regions have comparable total areas of mangrove habitat, but widely different numbers of genera and species (Table 10.1). Clearly mangrove species are not evenly distributed around the tropics. Many more species and genera are present in the IWP, particularly in the Australasian and Indo-Malesian subregions, than in the ACEP region.

The contrast between the two major regions is highlighted in Table 10.2. The IWP contains not just different species and even genera of mangroves,

Table 10.1 Area of mangrove habitat, and genera and species of mangrove, found in the different zones of the Indo-West Pacific (IWP) and Atlantic–Caribbean–East Pacific (ACEP) regions. From Ricklefs, R.E. and Latham, R.E. (1993). Global patterns of diversity in mangrove floras. In Species Diversity in Ecological Communities (ed. R.E. Ricklefs and D. Schluter), pp. 215–229. ©University of Chicago.

Region and subregion	Mangrove area (km²)	Genera	Species
IWP			
Australasia	17 000	16	35
Indo-Malesia	52 000	17	39
East Africa	5 000	8	9
ACEP			
West Africa	27 000	3	5
Western Atlantic/Caribbean	48 000	3	6
Eastern Pacific	19 000	4	7

Table 10.2 Number of genera, and of species, exclusive to the Atlantic–Caribbean–East Pacific (ACEP) and to the Indo-West Pacific (IWP) regions, and the number common to both regions. Data from Duke (1992).

	IWP only	Both	ACEP only
Genera	4	3	20
Species	11	1	57

but many more of them: more than three times as many genera (23 compared with seven), and nearly five times as many species (58 compared with 12). Depending on whether certain species are classified as true mangroves, or mangrove associates (see Chapter 1), these figures can be made to differ slightly, but the general conclusion is always that the IWP contains substantially more species and genera of mangrove than the ACEP (Ricklefs and Latham 1993).

Figure 10.2 compares the relative similarity of the subregions with respect to the actual species present. The six subregions fall into two distinct clusters, confirming the overall division into IWP and ACEP regions. No species are found in both ACEP and IWP regions, apart from the mangrove fern *Acrostichum aureum*. *Avicennia marina* (an IWP species) has recently been artificially introduced to California, and the mangrove palm *Nypa fruticans* to West Africa and to the Atlantic coast of Panama.

Only the two genera containing most species, *Avicennia* and *Rhizophora*, and the mangrove fern *Acrostichum*, occur naturally in both regions. Two sister genera, *Laguncularia* and *Lumnitzera*, probably separated relatively recently from a cosmopolitan ancestor: the former genus occurs only in the ACEP region, the latter in the IWP region (Duke 1992; Ricklefs and Latham 1993).

The number of species of mangrove-associated animals probably also tends to be significantly greater in the IWP, although this is much affected by collector bias. For some reason not every scientist is equally fascinated by the taxonomy of, say, polychaete annelids. As a result, the number of species recorded at a site depends heavily on whether a polychaete expert happens to have visited or not.

Among the most important groups of mangrove animal there are certainly more species of sesarmid crab in the IWP: at least 37 species have been recorded in Australia and 44 from Malaysia (and perhaps a further 20% still

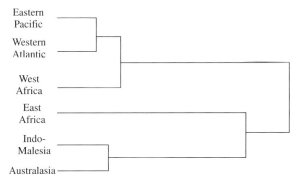

Fig. 10.2 Similarity of the species composition of the mangrove flora found in the six biogeographical regions referred to in the text. Horizontal distance indicates degree of similarity.

to be described), compared with eight or so from West Africa. Some 211 species of mollusc have been recorded from the Indo-Malesian area, and 145 from Australasia, compared with 124 from the West Atlantic/Caribbean. With due scepticism about some of the evidence, it can probably be concluded that IWP mangrove habitats are richer in at least crabs and molluscs, both key groups in mangrove ecology, although the effect is less marked than with mangrove trees themselves (Ricklefs and Latham 1993).

The origin of the disparity in species between the two major biogeographical regions is discussed in the next section.

10.2.2 Origins

How has the present distribution of mangrove species come about? One clue comes from the observation that more genera and species of mangrove occur in south-east Asia than elsewhere in the world. Figure 10.3, for instance, shows the geographical distribution of the four related mangrove genera of the family Rhizophoraceae, and similar distribution maps could be drawn for a number of other mangrove genera. A traditional explanation has been that the south-east Asian centre of diversity represents also a centre of speciation: that mangroves originated here and disseminated eastwards across the Pacific, to the western coast of the Americas, and westward to East Africa, and then to the east and west coasts of the Atlantic. The expansion of a species would be limited by the dispersal properties of

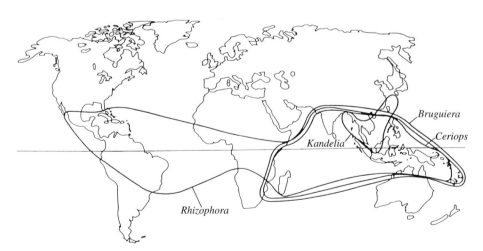

Fig. 10.3 Distribution of four mangrove genera: *Kandelia*, *Rhizophora*, *Ceriops*, and *Bruguiera*, of the family Rhizophoraceae. Reproduced with permission from Ricklefs, R.E. and Latham, R.E. 1993. Global patterns of diversity in mangrove floras. In *Species Diversity in Ecological Communities* (Ricklefs, R.E. and Schluter, D., eds), pp. 215–229. University of Chicago Press, Chicago. ©1993 by the University of Chicago Press.

its propagules, by the distribution of suitable habitats, and by major physical barriers. As not all species were equally good at dispersing, diversity would tend to decline with increasing distance from the source.

It is hard to reconcile an eastward spread across the Pacific to the Americas with the absence of mangroves from many of the Pacific islands (Figure 10.1), and equally hard to see how mangroves could have spread round the southern tips of either Africa or South America. In fact the current distribution of mangroves cannot be understood in relation only to the present configuration of the land masses of the world. The present can be explained only by the past: by the chronology of mangrove evolution, and by the movement of continents.

Mangroves have a long and conservative fossil history. There are many problems in interpreting fossil evidence. Apart from the possibility of misidentification of poorly preserved remains, fossil material can often be assigned only to a genus, or even family. Not all present-day members of the family Rhizophoraceae, for example, are mangroves, so fossil evidence of the existence of member of this family is not evidence for the existence of mangroves. Even identification in the fossil record of a genus all of whose modern members are mangroves is not conclusive. Ideally, corroborative evidence of a mangrove environment is desirable, such as fossil molluscs identifiable with modern mangrove residents. Conversely, of course, a fossil may not be recognized as mangrove if it belongs to a taxonomic group that is now extinct, or which has no modern mangrove representatives.

Fossils interpreted as mangroves have been reported from the early Cretaceous era (Figure 10.4), or even earlier, but most are doubtful. The earliest reliable mangrove fossil known is of fruit of the palm *Nypa*, from the early Paleocene. Pollen fossilizes well, and is distinctive enough for identification to genus level, sometimes even to species level. The oldest mangrove pollen known is that of *Nypa*, from the end of the Cretaceous period, around 69 million years ago, and the early Palaeocene. Palaeocene *Nypa* pollen has been found in North and West Africa and eastern Brazil, as well as in south-east Asia. Thereafter, *Nypa* is known from Eocene deposits in the Caribbean, South America, Africa and Europe, as well as Asia. By the Miocene it appears to have retracted to Asia, and currently, apart from deliberate reintroduction in the Caribbean and West Africa, *N. fruticans* is limited to Sri Lanka, south-east Asia, the Philippines, Australia, and New Guinea (Figure 10.5; Duke 1995; Plaziat 1995; Plaziat *et al.* 2001).

Other mangrove species show a broadly similar pattern. Fossils assigned to *Rhizophora, Ceriops, Bruguiera, Avicennia*, and *Pelliciera* have been discovered in English, French, and Spanish strata of the late Palaeocene, Eocene, and Miocene, in some cases with corroborative evidence such as traces of oysters having attached to prop roots (p. 100). In addition, two extinct genera, *Wetherellia* and *Palaeowetherellia*, were probably mangroves; they

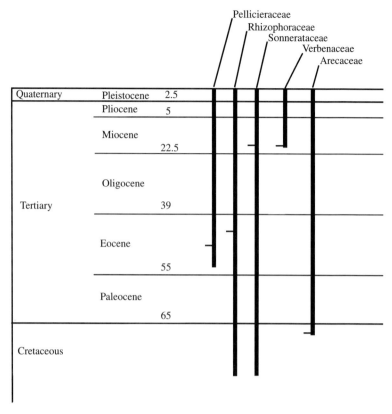

Fig. 10.4 The record of fossil pollen of five families containing mangrove species. The families include non-mangrove members: the horizontal bars indicate the oldest records of mangrove genera within each family. Figures are the approximate start of the geological period in question, in millions of years before the present. From Duke, N.C. 1995. Genetic diversity, distributional barriers and rafting continents—more thoughts on the evolution of mangroves. *Hydrobiologia* **295**: 167–181. Reproduced with kind permission from Springer Science and Business Media.

have been identified in marine Eocene deposits in Maryland, USA, in Germany, and in England, in association with *Nypa* (Ricklefs and Latham 1993; Plaziat 1995).

A parallel pattern of appearance and diversification is shown by molluscs belonging to groups known to be strongly associated with mangroves. This applies particularly to gastropods in the families Littorinidae and Ellobiidae. *Littoraria* (Littorinidae), for example, first appears in the Eocene in Europe, as a contemporary of fossil *Avicennia*, *Nypa*, *Pelliceria*, and *Rhizophora*. Thereafter, it broadly tracks the known distribution of fossil mangroves. Much the same is true of ellobiids and other gastropods with high fidelity to mangrove habitats (Ellison *et al.* 1999).

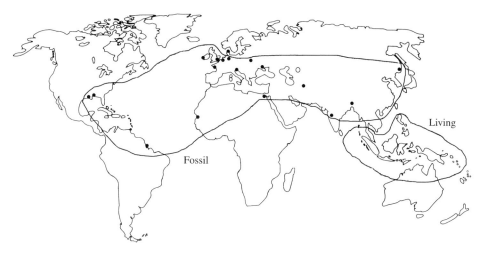

Fig. 10.5 Distribution of fossil evidence of the geographical distribution of the mangrove palm *Nypa*, compared with its current distribution (recent deliberate introductions excluded). Reproduced with permission from Ricklefs, R.E. and Latham, R.E. 1993. Global patterns of diversity in mangrove floras. In *Species Diversity in Ecological Communities* (Ricklefs, R.E. and Schluter, D., eds), pp. 215–229. University of Chicago Press, Chicago. ©1993 by the University of Chicago Press.

Such distribution patterns make no sense in relation to modern geography. At the time, however, Africa was separated from the land mass of Europe and Asia by a major sea, known as Tethys, which connected the Indian Ocean, through what is now the Mediterranean, to the forerunner of the modern Atlantic. North and South America were not joined at the Isthmus of Panama, so that the Pacific, the Tethys, and the Indian Ocean formed a continuous sequence. The present-day land barriers of North and South America, and Africa with Asia, did not exist, while America and Africa had not drifted far enough apart for the young Atlantic to limit dispersal. There was no impediment to the spread of mangroves along the margins of warm seashores, virtually around the world (Figure 10.6).

Thereafter, the latitudinal ranges of many species contracted, probably due to a general cooling of the Earth. In Europe and North Africa, mangroves disappeared completely; in Central America, some species dwindled while others vanished. Then, around 30–35 million years ago, closure of Tethys separated the IWP region from the Atlantic, and widening of the south Atlantic completed a barrier between the mangroves of the eastern and western Atlantic margins. North and South America united for the last time, with the Isthmus of Panama finally separating the eastern Pacific from the western Atlantic species. This took place a mere 2–3 million years ago, so that there are few differences in the mangrove floras on either side of the Isthmus (Duke 1995).

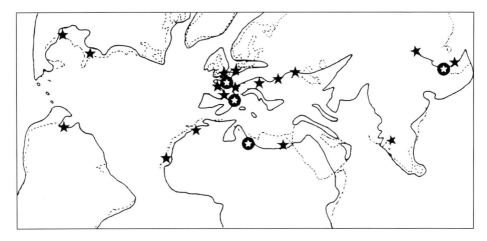

Fig. 10.6 Position of the major land masses during the Eocene period. The symbols indicate the locations of mangrove fossils from the period: black star, *Nypa*; star in circle, other mangrove species. Reproduced with permission from Plaziat, J.C. 1995. Modern and fossil mangroves and mangals: their climatic and biogeographic variability. *Geological Society Special Publication* **83**: 73–96, with permission of Geological Society and Jean–Claude Plaziat.

Tectonic movements are therefore responsible for the segmentation of the originally continuous mangrove distribution. Following the separation, regional extinctions and diversification of species proceeded. As the outcome is many more genera and species in the IWP than in the ACEP regions, one must conclude that fewer species became extinct, or that more evolved, in the IWP. It is virtually impossible to resolve the issue, but the balance of opinion is that the wet climates more widespread in the IWP offered more scope for terrestrial species to adapt to the mangrove habitat along a gradual salinity gradient. Arid climates—typical of Africa and the New World tropics at the periods of major speciation of mangroves—are intrinsically less favourable for mangroves, and tend to produce an abrupt transition with a zone of high salinity at the upper intertidal level of the shore. This might have inhibited the emergence of new mangrove species by extension of terrestrial species into the intertidal zone (Ricklefs and Latham 1993).

10.2.3 Local diversity

The species present at a specific location are a subset of the species present in the region as a whole. The pattern of local diversity of mangrove species quite closely follows regional trends. Not only does the ACEP region have fewer mangrove species overall, but locations within the region typically contain only three or four species. This represents about half of the total regional pool of species. As not every ACEP species is geographically widespread within the region, virtually all of the available species are represented at each location. This contrasts with the IWP, which has many more species overall and greater heterogeneity between sites in species present (Ricklefs and Latham 1993).

What determines the number of species at a particular location, as opposed to the species richness of an entire region? In general local processes such as competition tend to reduce species number: regional processes tend to increase local diversity through the movements of individuals and the resulting spread of species between habitats and habitat patches (Ricklefs and Schluter 1993).

Similar species at the same trophic level—such as the species of mangrove tree—compete for similar resources. Interspecific competition may have two outcomes. On the one hand, less successful competitors may simply disappear. Alternatively, competing species may minimize the cost of competition by becoming more specialized. With mangroves, this might manifest itself in the occupation of narrower segments of a salinity gradient, or of tidal level. Conversely, where an ecosystem contains fewer competing species, the outcome might be niche expansion, with species being more broadly distributed in relation to clinal environmental variables.

The ACEP has fewer species than the IWP, locally as well as regionally. If less intense interspecific competition favours broader ecological niches, ACEP species should, in general, show expanded niches: they should be found across a wider range of environmental variables than IWP species.

With this in mind, the distribution of ACEP and IWP mangrove species within sites can be compared. Table 10.3 shows where species are found with respect to two variables, height in relation to tidal range, and distance

Table 10.3 Occurrence of mangrove species at different estuarine locations and tidal levels for **Indo-West Pacific** (IWP) and **Atlantic–Caribbean–East Pacific** (ACEP) species. Figures in brackets are percentages of the total species in the region for which data were available. Estuarine locations were divided arbitrarily into upstream, intermediate, and downstream thirds of the estuary. From Duke, N.C. (1992): Mangrove floristics and biogeography. In Tropical Mangrove Ecosystems, ed. A.I. Robertson and D.M. Alongi, pp 63–100. ©The American Geophysical Union.

Tidal level	Estuarine location . . .	Occurrence of mangrove species		
		Downstream	Intermediate	Upstream
IWP				
High		11 (21)	20 (38)	9 (17)
Mid		15 (23)	23 (43)	9 (17)
Low		9 (17)	10 (19)	6 (11)
ACEP				
High		5 (42)	5 (42)	2 (17)
Mid		5 (42)	5 (42)	1 (8)
Low		2 (17)	1 (8)	0 (0)

up or down an estuary. Because the data were assembled from many published sources, where different classifications of tidal range were applied, tidal height is divided simply into high, middle, and low. This reflects inundation frequency, a key variable in mangrove distribution (Chapter 2). Estuarine position was divided into downstream, intermediate, and upstream. This determines the mix of river and sea water to which a tree will be exposed, and therefore reflects salinity exposure, another key environmental variable. The matrix combining the two therefore gives an impression of overall niche breadth, at least as far as these two environmental variables are concerned. In Table 10.3, progression from the upper right-hand corner (upstream in the estuary, high tidal position) to the lower left (downstream, low tidal position) can be regarded as a reflection of a shift towards more advanced mangrove habits, with greater tolerance of frequent inundation and higher salinity.

Mangrove species in the ACEP region do not appear to show broader ecological distributions. Indeed one sector—upstream, low tidal position—is apparently devoid of mangrove species, although in the IWP several species occupy this ecological position. Moreover, ACEP species are no more likely to spread into more than one of these ecological categories. In both regions, 42% of the species occur in only one of the nine possible combinations of estuary and tidal levels, and 58% in more than one (Duke 1992). There is, therefore, no clear evidence that processes such as interspecific competition play a major part in determining the number of mangrove species found at a particular location.

Stochastic factors may also affect the number of species of mangrove present. It is well known that the number of species present on an island is often related to the area of the island (Rosenzweig 1995). An island, in this context, can be any area of suitable habitat separated from other such areas by stretches of unsuitable habitat: it need not necessarily be a physical island surrounded by sea. The species–area relationship results largely from a balance between the successful establishment of new colonizers and the chance extinction of established species, both being affected, differently, by the available area. With mangrove species, establishing any such relationship is complicated by many other factors, such as defining islands. In some cases, however, it is possible to see how area might locally affect species number. For a group of four adjacent West African islands, Bioko (Fernando Po), Principe, Saō Tomé, and Annobo, mangrove species numbers correlate closely with island area (see Figure 10.7). Similar species–area relationships apply at all scales, including the comparison of total mangrove area and species number between the major biogeographical regions (Ellison et al. 1999; Ellison 2002).

To be present on an island, a species must first get there: floating propagules must have arrived in sufficient numbers to establish a founding population. The longer the time at sea, or the greater the distance from source, the lower

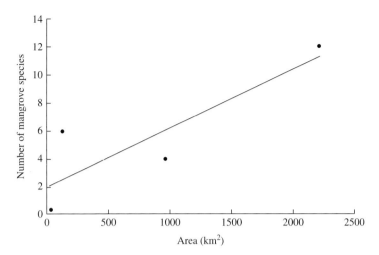

Fig. 10.7 Mangrove species number in relation to area (km^2) for the four West African islands of Bioko (Fernando Po), Principe, Saõ Tomé, and Annobon. Reproduced with permission from Saenger, P. and Bellan, M.F. 1995. *The mangrove Vegetation of the Atlantic Coast of Africa. A Review*. pp. 1–96. Laboratoire d'Ecologie Terrestre, Toulouse.

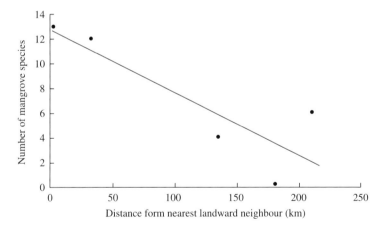

Fig. 10.8 Mangrove species number in relation to distance from the mainland for the four West African islands of Bioko, Principe, Saõ Tomé, and Annobon, and for Cameroon. From data in Saenger, P. and Bellan, M.F. 1995. *The mangrove Vegetation of the Atlantic Coast of Africa. A Review*. pp. 1–96. Laboratoire d'Ecologie Terrestre, Toulouse. With permission.

the chances of success. Figure 10.8 shows the relationship between mangrove species number and distance from the mainland of Africa to each of the same four islands. For comparison, the nearest country of the African mainland, Cameroon, has been added to the graph. Species number is clearly related, inversely, to distance from the mainland. (A similar relationship appears if distance from the nearest landward island is plotted instead of distance from the mainland, on the assumption that mangroves might have used each successive island as a stepping stone to the next.) This relationship is much the same as that between mangrove species on various islands of the Pacific and distance from Australia or New Guinea (Figure 10.1), but at a distinctly smaller scale.

The species most widely dispersed across these islands should be those whose propagules last longest at sea. In general, this appears to be the case. *Avicennia germinans* and *Rhizophora harrisonii*, with propagule-survival times estimated at 3 months and more than a year, respectively, are found on all three of the mangrove-occupied islands (as well as on the adjacent mainland). *Laguncularia racemosa*, which survives only a month at sea (Table 2.5), is restricted to the island nearest to the mainland, Bioko. Surprisingly, however, the two other species of *Rhizophora*, *R. mangle* and *Rhizophora racemosa*, are restricted to the African mainland and the nearest island, Bioko. As far as is known the propagules of these species have similar dispersal abilities to those of *R. harrisonii*.

10.2.4 Genetic diversity

Biodiversity can also be considered at a level below that of the species. The extent of genetic differences between individuals within a population is an important measure of the diversity of that population. Assessment of the genetic differences between populations of the same species gives a measure of the extent to which such populations are genetically isolated from each other, whereas genetic comparison of different species can give information about the extent of divergence and, potentially, when such divergence took place. The genetic continuities and discontinuities of *Rhizophora* populations, and their relationship to distribution patterns, have recently been reviewed by Duke *et al.* (2002).

Among the red mangrove of Florida and the Bahamas (*R. mangle*), studies of chlorophyll-deficient mutations indicate that the population relies predominantly on self-pollination: more than 95% of embryos appear to result from self-crossing. As a result of inbreeding, the populations contain relatively little genetic variation. *Rhizophora* tends to form dense monospecific forests. Agricultural ecosystems are also genetically uniform monocultures and, as a result, are often particularly prone to disease and pest infestations. This suggests that the low species and genetic diversity of mangroves might be a factor in the occasional mass-defoliation episodes which have been recorded (p. 75).

However, such high rates of self-pollination, and low genetic diversity, are not universal. The same species of mangrove in Puerto Rico had a lower rate of self-pollination (71.2%), so the mode of pollination adapts to local circumstances rather than being a fixed species-specific characteristic. Populations with more frequent self-pollination had higher mutation rates, and one intriguing possibility is that an increased mutation rate has evolved to offset the low genetic variation resulting from inbreeding (Klekowski et al. 1994).

DNA polymorphism can now be assessed directly, so that genetic analysis is no longer restricted by the luck of finding appropriate mutations (and the difficulty of carrying out significant breeding programmes). Molecular genetics studies on *Avicennia* in India have been used to assess and compare the amount of genetic variation within separate populations, between populations of the same species, and between species. *Between* populations, genetic variation is relatively high, suggesting that the separate populations are largely isolated from each other, with little mutual gene flow. The amount of genetic variation *within* a population varies from one population to another, suggesting that it is dictated by local circumstances. Finally, comparisons between species of *Avicennia* (*A. marina*, *Avicennia alba*, and *Avicennia officinalis*) indicate that *A. marina* is closer to *A. alba* than to *A. officinalis*, and presumably diverged from it more recently (Parani et al. 1997).

These examples give some idea of the potential of genetical analysis in understanding the population biology of mangroves. Genetic studies of mangroves are in their infancy, but have already provided some useful information and promise much more, even if so far they have raised more questions than have been answered.

10.2.5 Diversity of the mangrove fauna

Local species richness of the mangrove fauna is also affected by factors such as habitat area, dispersal ability, and distance from areas of similar habitat. By far the most detailed studies are the classical experiments of Simberloff and Wilson on some small mangrove islands among the Florida Keys (Simberloff and Wilson 1969, 1970; Simberloff 1976).

First, they established an inventory of the species of terrestrial invertebrate (principally insects) present on each of a series of small mangrove islets, which ranged from 11 to 25 m in diameter. The islets were then encased in plastic sheeting, and fumigated with methyl bromide, killing all terrestrial arthropods present. Not, perhaps, the most environment-friendly of experiments: however, the fumigated islets were extremely small (some little more than a single tree) and recovered rapidly.

Periodically, over the ensuing 2 years, the recolonizing species present were recorded. In each case, the number of species rose, then levelled off, roughly asymptotically, at a number of species close to the number present before

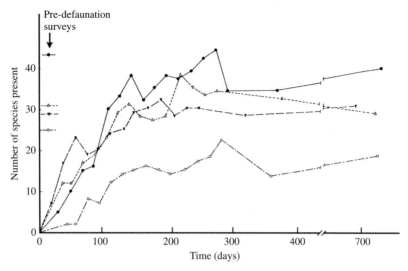

Fig. 10.9 Number of species of terrestrial arthropod present following fumigation of four mangrove islets in the Florida Keys. The number of species identified at each islet before fumigation are indicated on the vertical axis. The islets were similar in area, but differed in distance from the nearest possible source of immigrants. The islet with the lowest number of species before fumigation, and after recovery, was farthest from a source, and that with the greatest number of species before and after the experiment was closest to a source. Reproduced with permission from Simberloff, D.S. and Wilson, E.O. 1970. Experimental zoogeography of islands. A two year record of colonization. *Ecology* **51**: 934–937. ©Ecological Society of America.

the fumigation (Figure 10.9). However, although the total number of species present was about the same as that before the fumigation, the actual species present were not the same ones as before; nor were they necessarily the same from one census to the next. As an example, seven species of Hymenoptera (ants, bees, and wasps) were present on one islet before fumigation, and eight species a year after: but only two of the species were present on both occasions. Species were therefore continually being lost, and gained, and the total present at any one time was a snapshot of a dynamic equilibrium.

What determines the equilibrium species richness for one of these mangrove islets? As with tree diversity, discussed above, the number of species of terrestrial invertebrate on Florida Keys mangrove islands correlates inversely with distance from source habitats, and directly with the island area.

The problem with simply correlating area with species number is that larger areas may incorporate greater habitat diversity than smaller ones. A close species–area relationship does not allow the inference that area determines species number, since a high level of diversity might result from greater

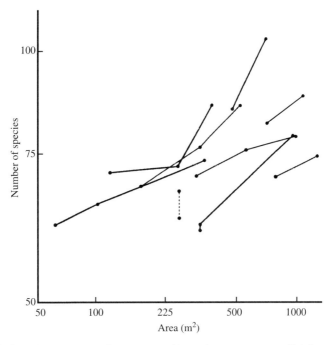

Fig. 10.10 Individual species/area curves for mangrove islets whose area was artificially reduced. Points relating to each islet are joined. The islet indicated by two points joined by a dashed line was a control whose area was not altered, and the change in species number presumably reflects random change. From Simberloff, D. 1976. Experimental zoogeography of islands: effects of island size. *Ecology* **57**: 629–648. ©Ecological Society of America.

habitat complexity, rather than from greater area. Simberloff and Wilson ruled out this possible confusion of variables by artificially reducing the area of a group of mangrove islets with power saws. This had the effect of reducing area without any concomitant change in habitat complexity of the surviving vegetation. The result was that invertebrate species diversity fell, demonstrating its dependence on habitat area rather than complexity (Figure 10.10).

Terrestrial arthropods and isolated mangrove islets obviously constitute a special case. Most mangroves are much less isolated from sources of colonizing species, either terrestrial hinterland or the sea. With larger and longer-lived species than the insects and other terrestrial arthropods, and with larger areas of mangrove than these tiny islets, the dynamics of the species turnover will in any case be vastly different. Nevertheless, Simberloff and Wilson's experiments make the crucial point that the species composition and richness of the mangal is not necessarily static, or the result of wholly deterministic processes.

10.3 Seagrass biogeography and biodiversity

The geographical distribution of seagrasses is similar to that of mangroves, with the exception that seagrasses extend into temperate seas. The greatest species richness occurs in south-east Asia, with species numbers declining with increasing latitude, and with longitudinal distance (see Figures 1.2 and 1.3; Duarte 2001).

As with mangroves, the modern pattern of species distribution is constrained by environmental requirements of particular species; but also by past events and processes. In particular, the tectonic reconfiguration of the Earth's landmasses is the key to explaining the modern distributions of species. Dispersal abilities of different seagrass species have also modified the pattern.

Direct information about the evolution of seagrasses and their past distribution comes from fossil evidence. The fossil record of seagrasses is not very extensive, but sufficient to establish, in broad outline, what probably happened. The first angiosperms returned to the sea probably around 100 million years ago, in the Cretaceous period. Cretaceous seagrasses are poorly known. Good fossils of two genera, *Thalassocharis* and *Archaeozostera*, are known from Japan and western Europe, respectively, whereas the modern genus *Posidonia* is represented by a rather poorly known species, *Posidonia cretacea*. By the late Eocene, about 40 million years ago, most modern genera of seagrasses had evolved, including *Thalassia*, *Thalassodendron*, *Cymodocea*, and *Halodule*, whereas *Enhalus* and *Phyllospadix* appeared more recently (Larkum and den Hartog 1989).

The origins of seagrasses are unclear from the fossil evidence, but molecular comparisons of modern species support the view that they are polyphyletic—that the seagrass mode of life evolved more than once, independently. *Halophila*, *Enhalus*, and *Thalassia* (family Hydrocharitaceae) most probably evolved from freshwater ancestors that acquired salinity tolerance and spread down watercourses into the sea; while the remaining species (family Potamogetonaceae) evolved separately from either aquatic or salt-marsh ancestors (Les *et al.* 1997).

The initial invasion of the sea by ancestors of the modern seagrasses occurred along the shores of Tethys, in parallel with the evolution of mangroves (Figure 10.6). Thereafter, the pattern of species distribution is largely explained by tectonic movements, the imposition of barriers segmenting the Tethyan distribution into regions, and subsequent divergence. Barriers explain, for example, the occurrence of twin species of *Thalassia*, *Halodule*, and *Syringodium* in the Pacific and Caribbean, and of *Zostera* and *Phyllospadix* on opposite sides of the Pacific, separated, respectively, by the Isthmus of Panama and the vast area of the central Pacific (Table 10.4; Duarte 2001).

Throughout their history, seagrasses seem never to have diversified greatly. Some species known from the fossil record have become extinct, but there

Table 10.4 Examples of closely related species of seagrass, diverging after separation by a physical barrier. From Duarte (2001).

Indo-Pacific	Caribbean / West Atlantic
Thalassia hemprichii	*Thalassia testudinum*
Halodule uninervis	*Halodule wrightii*
Halodule pinifolia	
Syringodium isoetifolium	*Syringodium filiforme*
Temperate west Pacific	Temperate east Pacific
Zostera caulescens	*Zostera marina*
Zostera asiatica	
Zostera caulescens	
Zostera japonica	
Phyllospadix iwatensis	*Phyllospadix scouleri*
Phyllospadix japonicus	*Phyllospadix serrulatus*

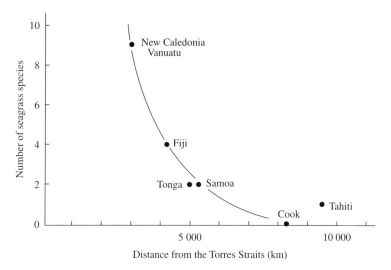

Fig. 10.11 Relationship between species richness of seagrass meadows and distance from the Torres Strait. From Mukai, H. (1993). Biogeography of the tropical seagrasses in the western Pacific. Australian Journal of Marine and Freshwater Research **44**, 1–17. ©CSIRO Publishing.

seem never to have been many more seagrass species than the 50–60 there are at present, out of many hundreds of thousands of known angiosperms. This is probably a consequence of the reproductive constraints of a marine life and the tendency of seagrasses to spread vegetatively rather than by sexual reproduction (p. 46; Ackerman 1998).

Seagrass distributions have probably also been affected by dispersal ability. The greatest species richness of seagrasses is in south-east Asia: species numbers decline with distance from this region, along the trajectory of the principal ocean currents. Figure 10.11 shows the decline in species along the Equatorial Current which runs eastward from Torres Strait at the northern tip of Australia (Mukai 1993). Dispersal may be as floating seeds, or as viable fragments of leaf or rhizome.

10.4 Diversity and ecosystem function

It seems reasonable to suppose that the species richness of an ecosystem will have some bearing on its function. In particular, it has been postulated that a diverse ecosystem will be more productive, or more resilient, than a species-poor one. The precise reasons for these expectations need not concern us, but clearly the predictions could be tested.

Real ecosystems are generally too complex, and subject to too many influences, to give a clear answer. Ecologists have therefore resorted to experiments with artificial ecosystems, with contrived levels of diversity and mixtures of species. These have included systems as different as grassland plots, small-scale ecosystems set up in artificial microcosm containers, and assemblages of micro-organisms (Tilman 1996; McGrady-Steed *et al.* 1997; Naeem and Li 1997).

Although the results are complex, some general conclusions have emerged. Adding more species tends to increase productivity of the system as a whole—but only up to a certain limit. Further increases in species number have a greater effect on ecosystem stability: the more species present, the smaller the fluctuations with time, and differences between replicates, in measures of ecosystem function as a whole.

What initially increases productivity is probably the inclusion of additional functional groups of organisms. (A functional group is a particular mode of exploiting resources within the environment. Trees, shrubs, and epiphytic algae, for instance, all operate at the same trophic level, but in very different ways.) Once an appropriate mix of functional groups has been reached, adding further species may intensify competitive interactions between species, but will not significantly enhance productivity of the ecosystem as a whole. A further increase in species richness may increase stability and predictability of the ecosystem without a concomitant increase in productivity.

Natural ecosystems are generally so rich in species, and affected in such complex ways by environmental variables, that there is little scope for testing the relationship between diversity, productivity, and stability. Mangrove and seagrass ecosystems may be simple enough to address the question of the functional significance of diversity. Although the mangrove fauna is relatively rich in species, the habitat is generally dominated by a small number of plant species, and is structurally simple compared with other tropical forests. Moreover, mangrove forests occur in similar environmental settings in many different parts of the tropics, but the number of plant species varies greatly between different biogeographical regions. Much the same is true of seagrass meadows. It is therefore possible to find comparable mangrove or seagrass ecosystems, in similar environmental conditions, which differ principally in plant species richness, and to test whether any differences in ecosystem function can be related to differences in diversity.

If productivity was directly related to species richness, mangrove habitats of the IWP should be more productive than those of the ACEP. This appears not to be the case. Productivity and standing-crop biomass vary considerably between sites, and tend to decrease with increasing latitude, but a comprehensive collection of the available data shows no particular differences between biogeographical regions (Saenger and Snedaker 1993). Such a general comparison masks local differences in many factors known to affect productivity, such as environmental setting and salinity gradients. A better experiment would be to compare the productivity of mangroves in a large-scale transect from west to east across the islands of the Pacific, where climate and substrate are similar but species number, limited by dispersal ability, varies from 30 in New Guinea to a single species in some of the more outlying islands (Figure 10.1; Woodroffe 1987).

Mangroves tend to arrange themselves into relatively homogeneous, almost monospecific patches or bands, often aligned with physical gradients of the environment (Chapter 4). At a scale of hectares, a mangrove forest may encompass many species and be regarded as relatively diverse. At a scale of square metres, in contrast, diversity appears low, and a sample may include only a single species. Tree diversity may not appear to correlate with productivity because at the spatial scale relevant for species interactions the species scarcely intermingle.

In one case, the spatial scales of diversity and productivity measures are compatible. This is the 'experiment' of the Matang, a managed mangrove ecosystem in western peninsular Malaysia, described more fully in the next chapter (p. 211). One effect of the management régime, over the last 90 years or so, has been to reduce tree diversity, virtually to a single species. The productivity appears to be declining. A clear-felled patch of forest in the 1970s yielded 136 t/ha, whereas a decade previously the yield had been 158 t/ha. However, there is no particular reason to attribute this apparent decline in productivity to a reduction in tree diversity, rather than to other factors such as the effect of invading *Acrostichum* reducing seedling growth (Chapter 11), or to alterations in management practice (Gong and Ong 1995).

Adult mangrove trees may not always interact much in life, but their principal contributions to ecosystem processes, shed leaves, intermingle on the forest floor. Mixed-species leaf litter decomposes no faster than leaves from a single species, suggesting again that tree diversity has little effect on the rate of energy flow through the ecosystem.

Litter processing could also be affected by the diversity of either leaf-processing crabs (p. 103), or of decomposing micro-organisms. Crab abundance certainly affects litter-processing rates, but the impact of crab diversity has not been investigated. Microbial diversity is a problematic concept. Diversity is generally extremely high, with many species globally distributed: 'in the case of micro-organisms, "everything is everywhere" ' (Fenchel et al. 1997). In natural circumstances, microbial biodiversity probably does not vary sufficiently to affect mangrove ecosystem processes.

With the possible exception of the Matang, there is little evidence that variations in biodiversity—whether of mangrove trees, the mangrove fauna, or micro-organisms—has any significant effects on ecosystem productivity.

Whether or not it affects productivity, tree diversity may also be relevant to the long-term stability of a mangrove forest. In many parts of the world, mangroves are subjected to intermittent catastrophic damage from typhoons or hurricanes. These create gaps, often filled by different species from the ones that have been destroyed. A forest with a mix of species, some more resistant to hurricane damage, and others better at rapidly occupying gaps, may be more stable over a timescale of decades. Here, too, the significant factor may be functional differences between species, rather than actual species diversity. Species-rich IWP mangrove forests should recover more rapidly than relatively species-poor ones in the ACEP region, but it is difficult to arrange for comparable forests to be visited by appropriately matched hurricanes in the two regions.

It is easier to investigate the relations between diversity and ecosystem function in seagrasses: compared with mangroves they are smaller, more amenable to experimental manipulation, and operate on a timescale more compatible with the duration of research funding and the lifespan of scientists.

In general, species richness of seagrasses correlates with higher biomass and production. However, this effect appears to be related not to species richness as such, but to the variability within the seagrass meadow of the size of its constituent species. Larger species function in some respects differently from smaller ones, and the functional complementarity of species of different sizes may serve to enhance the function of the community as a whole. This may include a direct contribution to production by a relatively small understorey seagrass, or to community stability by the presence of a small gap-colonizing species which would minimize the loss to overall community production that would otherwise follow disturbance (Hemminga and Duarte 2000).

Many seagrass beds are virtually or entirely monospecific, and many seagrass species reproduce almost entirely clonally. Species diversity is negligible, and intraspecific genetic diversity low. Nevertheless, genetic diversity within a species may still be relevant to ecosystem function.

Temperate populations of *Zostera marina* typically include a number of genetically different clones. Two recent experimental studies have demonstrated the importance of genetic diversity to the ability of seagrass meadows to recover from disturbance. In the first, experimental plots in Bodega Bay, California, were planted out with different mixes of *Zostera* clones: one, two, four, or eight genotypes. Shoot density was then monitored over the following year, including a period when a large number of Brent geese (*Branta bernicla*) migrated into the area and grazed intensively on the *Zostera*. Survival and recovery from this onslaught was broadly proportional to genetic diversity within an experimental plot (Hughes and Stachowicz 2004).

The second study, in the Baltic Sea, similarly involved planting out experimental plots of *Z. marina*—in this case, one, three, or five genotypes—and subsequent measurement of biomass and shoot density. Fortuitously, this experiment was initiated shortly before an unprecedented heat wave, with temperatures frequently exceeding 20°C, at which *Zostera* stops growing, and even 25°C, at which it starts dying off. It therefore became an investigation of the resilience of *Zostera* meadows to severe environmental stress. Again, the genetically more diverse plots survived and recovered better, with higher shoot density and biomass. The invertebrate fauna also benefited, with greater biomass (but not diversity) of detritivores, filter feeders, and grazers all recorded in the more genetically diverse plots (Reusch *et al.* 2005).

It is relatively easy to conceive small-scale experiments, over short timescales, to test relatively straightforward predictions about the relationships between biodiversity and ecosystem functions and resilience. If larger-scale factors are taken into account, such as habitat fragmentation and dispersal between patches, predictions from theory become more complex and the effective testing of these predictions more difficult. At larger spatial scales, and longer timescales, experimental tests become more difficult. However, some interesting recent studies indicate that, at a larger scale, increasing diversity does not necessarily buffer a seagrass ecosystem against perturbations. Our understanding of the functional significance of biodiversity in seagrass communities is still in its infancy (Duffy 2006, France and Duffy 2006).

11 Impacts

Humans benefit in many ways from mangrove ecosystems, deriving direct products such as timber and fuelwood, and services such as support of fisheries and protection against coastal erosion. Direct harvesting of seagrass is much less significant, but seagrasses are nevertheless also the source of important goods and services.

11.1 Mangroves

Inevitably, significant exploitation of mangroves means that few mangrove forests are now pristine: most have been affected to some degree. Any extraction removes material which would otherwise have contributed to the ecosystem. Selective removal of certain size categories of tree for timber affects the age structure and architecture of the forest, and selective removal of some species in preference to others affects the community structure and reduces diversity. Physical disturbance during the extraction process may prevent the establishment of seedlings to replace the trees removed.

Large-scale exploitation of a mangrove forest can be carried out sustainably so that the benefits can be maintained indefinitely. The price of sustainability may be a somewhat simplified ecosystem, with lower biodiversity and possibly a decline in productivity, but with the major ecosystem functions intact. A particularly good example, the Matang mangroves of Malaysia, is discussed below (p. 211).

Properly conceived and enforced sustainable management systems are depressingly rare. In too many cases exploitation is completely unregulated, a management system is inadequate, or enforcement is lax. The result of overexploitation is habitat degradation, reduction in area, and ultimate loss of the resource being exploited. Mangrove loss may also be followed by increased erosion of shorelines, a decline in fish catches, or other unforeseen (but hardly unforeseeable) effects.

Unwise exploitation is not the only threat. The benefits of mangroves are often undervalued, or simply not recognised, and mangrove habitats regarded as mere waste land, suitable only for conversion to an alternative use seen as more beneficial. Mangroves are the subject of land 'reclamation' for building purposes, or for conversion to agricultural land. (Even though subsequent erosion may reduce the value of the land for building, or a saline and anoxic soil may prove less than ideal for growing crops.) A particularly widespread and serious impact has been the clearance of mangroves for the construction of shrimp ponds.

Conversion of mangrove land to other uses may be justifiable if the gains are greater than the losses, but only rarely have attempts been made to evaluate these fully. Economic analysis of the true value of the products and services supplied by mangroves is likely to prove essential in future rational management. Ecologists may prefer to measure productivity in Joules rather than in dollars, but the latter units have more impact on decision-makers.

Mangroves additionally face indirect and accidental threats. Alteration of a hydrological régime by the extraction of river water for irrigation, many miles from the sea, may alter estuarine salinity and sedimentation. This has certainly been a factor in the reduction of mangroves of the Indus delta in Pakistan (p. 222). Florida offers another example of inadvertent damage to coastal mangrove forests caused by changes in terrestrial water use. Pollution, particularly by oil, can also cause severe local damage to mangrove forests. Not all losses are attributable to human actions, however. Natural hazards, such as typhoons and hurricanes, from time to time cause serious local damage.

Sometimes losses have been mitigated, or even reversed, by habitat restoration and mangrove-replanting programmes. Occasionally, just as human activities have inadvertently destroyed mangroves, they may equally unintentionally result in extensions of a mangal. Overall, however, the net effect of the various human impacts has been a drastic overall reduction in mangrove area, although this is difficult to quantify. Some figures may highlight the extent of losses in south-east Asia. In the Philippines around 60% of the original mangrove area has disappeared. In Vietnam, losses are estimated at 37%. In Thailand, 55% was lost between 1961 and 1986 (a year-to-year fall of around 3%). Between 1980 and 1990 Malaysia lost 12%, equivalent to 1.3% annually. The rate of loss varies greatly both between and within countries, but worldwide about 35% of total mangrove area has been lost over the last 20 years, and 2% of the surviving area of mangrove forest is being lost each year (Spalding *et al.* 1997; Valiela *et al.* 2001; Alongi 2002).

Looking beyond the immediate future, mangroves face further threats from predicted changes in climate, due at least in part to human activities: increased atmospheric carbon dioxide concentrations, global warming, and concomitant rises in sea level. All of these have long-term implications for mangrove ecosystems.

An attempt to catalogue all interactions between humans and mangroves would be impractical, as well as tedious. Humans exploit mangrove ecosystems in a multitude of ways, and have many and varied impacts upon them. This chapter therefore discusses only selected examples of some of the major ways in which mangroves and humans interact.

11.1.1 Uses of mangroves

The benefits of mangroves to humankind vary greatly between areas. Fringe mangroves (p. 50) may contribute relatively little productivity, hence may not be of much significance as a source of wood, or in contributing to fisheries. They are often of great importance in coastal protection. Riverine forests such as those of the Sundarbans of the Ganges delta are extensive in area, generally highly productive, and the basis of many products that can be extracted for human use. The precise goods and services available depend on the nature of the forest (Ewel *et al.* 1998).

11.1.1.1 *Direct uses*

Accurate information is generally available only for products which are extracted and marketed on a commercial basis. Local subsistence use is less well documented and probably undervalued, and knowledge of it is often little more than anecdotal. It can nevertheless be substantial and important. Some of the uses of mangrove products are listed in Table 11.1.

Mangrove wood is used in the construction of buildings, as firewood, and in the manufacture of charcoal (p. 211). Branches are used to make fishing poles and fish traps. In some areas wood extraction is a local industry, managed more or less sustainably in such a way as to benefit local people. In other instances, this is far from being the case. Vast areas of mangrove in

Table 11.1 Some uses for mangrove products. Mainly from FAO (1994).

Fuel	Charcoal, firewood
Construction	Timber, scaffolds, railway sleepers, mining props, boat building, dock pilings, beams and poles, thatch, matting, fence posts, chipboard
Fishing	Fishing stakes, fishing boats, wood for smoking fish, tanning for nets/lines, fish poison, fish-attracting shelters
Textiles	Synthetic fibres (rayon), dyes, tannin for preserving leather
Food, drink	Sugar, alcohol, cooking oil, vinegar, tea substitute, fermented drinks, dessert topping, seasoning (bark), sweetmeats (propagules), vegetables (fruit, leaves)
Domestic	Glue, hairdressing oil, tool handles, musical instruments, rice mortar, toys, matchsticks, incense, cigarette wrappers, cosmetics
Agricultural	Fodder
Medical	Treatment of ringworm, mange, toothache, leprosy, sore throat, constipation, dysentery, diarrhoea, boils, bleeding lice, fungal infections, bleeding, fever, catarrh, kidney stone, gonorrhoea, etc.
Miscellaneous	Paper manufacture

south-east Asia have been cleared to support the international woodchip industry, with no attempt at sustainable management (Ong 1995).

Mangrove fruits may be used as food, leaves as fodder for domestic animals (p. 225), and the bark is a source of tannins and in medicine (Bandaranayake 1998). (The medicinal use of mangrove materials may sometimes be overstated, however. When one group of West Africans was asked about the medicinal uses of mangrove products, they reeled off an impressive list. When asked instead 'what do you do when you are sick?', the answer was 'I go to the clinic'.)

The most versatile mangrove tree must surely be the Nypa palm (*Nypa fruticans*) of Asia, Australia, and (following deliberate introduction) West Africa. Nypa leaves are harvested in the construction of dwellings. The mature leaves are dried and the midribs removed; the fronds are then folded round a bamboo rod and stitched in place to make a shingle for roofing or walls. Leaf petioles are used as floats for fish nets, or chopped and boiled to obtain salt. Young leaflets are used as cigarette wrappers, and older ones to weave hats, umbrellas, raincoats, baskets, mats, and bags. The gelatinous endosperm of young seeds is eaten raw or preserved in syrup, or to flavour ice cream, while the ivory-like hardened endosperm of mature fruits is used to make buttons. Various components are used medicinally, in various parts of south-east Asia, to treat toothache, headaches, and herpes.

The sap, 14–17% sucrose, is a particularly valuable commodity. It is used to produce a soft drink and, after fermentation, a potent alcoholic beverage. (The medical use of Nypa products in the treatment of headaches may therefore seem particularly apposite.) Attempts have been made to develop Nypa alcohol production for industrial use. Yield is said to be increased by physical stimulation of the stalk. Methods include regular kicking or beating with a mallet, shaking the fruit to and fro, or a complicated routine of bending the stalk 12 times in one direction, patting it backwards and forwards 64 times, kicking its base four times, and repeating the whole process four times per week. These elaborate procedures apparently increase the daily sap yield from 155 to 1300 ml (Hamilton and Murphy 1988).

Not many mangrove species can boast quite such a wide and varied array of products, but most are of at least local significance in contribution to the economy.

11.1.1.2 *Indirect uses*

Mangrove animals are also an important source of food and other products. As with the direct extraction of mangrove wood, leaves, and fruit, the nature and extent of these varies from place to place. In many parts of the Indo-Pacific the larger mangrove-dependent molluscs, such as *Telescopium*, are collected for food or as a source of lime (see Chapter 6). Bivalve molluscs may also be extracted from the mangrove mud. Honey production may depend on bees using mangrove flowers and, of course, various commercial fisheries depend to varying extents on mangroves.

The dependence of fish and shrimp fisheries on mangroves has been discussed in Chapter 9. The extent of the linkage with fisheries is difficult to establish, partly for the reasons already discussed, and partly because only commercial fisheries are reasonably well documented. Subsistence fishing, which does not pass through commercial channels, can be of crucial importance to a local community. A good example of the diversity of mangrove-based fisheries is shown by Sarawak, on the northern coast of Borneo. A wide range of trapping techniques have evolved, to match the diversity of creeks and inlets in the mangal. More than 30 species of fish are caught, about 10 species of shrimp, the Mangrove crab (*Scylla* sp.), and two species of jellyfish (Pang 1989).

11.1.1.3 *Coastal protection*

Mangrove roots tend to retain sediment and consolidate the soil, and hence facilitate accretion and retard coastal erosion. In the Jiulong Jiang estuary, southern China, for instance, there has been a practice for many years of constructing earth embankments across the mouths of inlets and converting the area behind the barrier to shrimp ponds, or rice or lotus fields. In the process, the mangroves that fringed the shores of such inlets were often destroyed. Serious problems arose because the region is affected by typhoons which regularly destroyed the impounding barriers. Replacing the earth embankments with concrete, and even sinking a retired warship as a breakwater, simply showed that human-made materials were less effective against typhoons than earth embankments. Only when the local villagers were involved in a scheme to plant mangrove trees (*Kandelia*) systematically along the seaward side of earth embankments did they (and the shrimp ponds and paddy fields) survive. Mangrove roots compare favourably with concrete and steel as mechanical structures, besides, of course, contributing an exploitable source of wood, and molluscs and crabs for food.

A more dramatic test of the ability of mangroves to protect the coastline came with the Indian Ocean tsunami of December 2004, which travelled at more than 600 km/h, and struck coastlines as a wave of more than 15 m. Because coasts vary in topography, and because in most areas little precise information was recorded of the magnitude of the tsunami, it is difficult to quantify the protective effects of mangroves. Nevertheless, studies in India concluded that mangroves saved lives and mitigated damage: 'villages on the coast were completely destroyed, whereas those behind the mangrove suffered no destruction even though the waves damaged areas unshielded by vegetation north and south of these villages' (Danielsen *et al.* 2005).

11.1.1.4 *Ecotourism*

Mangroves can also be exploited for ecotourism. In Trinidad, tourists are attracted by the scarlet ibises (*Eudocinus ruber*) of Caroni Swamp (p. 89), and in Florida and Honduras by the chance to kayak among the mangroves. The nature reserve of Kuala Selangor, in Malaysia, is a particularly good example of well-organized ecotourism. It lies on the estuary of the river Selangor, within easy reach of the capital, Kuala Lumpur. The reserve area comprises an assortment of pools among the mangroves with pathways and hides from which visitors can watch kingfishers, herons, bee-eaters, and numerous other spectacular birds. Walkways have been constructed through the mangroves above high-tide level, so that it is possible to observe monitor lizards and crabs in comfort, and without the necessity of even setting foot in the mud. A few kilometres away, the famous synchronously flashing fireflies (p. 80) can be observed from small boats.

Although ecotourism has not yet been widely developed, it does represent a significant potential source of revenue, particularly where mangrove areas are close to centres of population, or to other tourist attractions.

11.1.2 Sustainable management: the case of the Matang

Very few mangrove forests are pristine: most are to some degree affected by human activities. If exploitation is uncontrolled, the result is often deterioration, loss of biodiversity, reduction in extent of the exploited forest, and drastic reduction of the resource being exploited.

In a few cases, an effective management régime has been put in place, such that intense exploitation is maintained without significant long-term deterioration. One of the best examples is that of the Matang forest of western peninsular Malaysia, which has been managed on a sustainable basis for more than a century. The managed area of the Matang consists of an estuarine complex of streams, creeks and inlets, amounting to more than 40 000 ha. Some 2000 ha are left untouched as so-called virgin jungle reserve, whereas further patches are set aside for research, or protected for archaeology, ecotourism, education, or as bird-sanctuary forests.

The principal harvest is wood for charcoal, a major domestic fuel. The management routine has been modified since its inception, and currently operates on the basis of a 30-year rotation. The forest is divided into blocks of a few hectares. Blocks are allocated to charcoal companies by the Forestry Department, who regulate the whole operation. Each block is clear-felled: workers simply move in by boat, demolish every tree with chainsaws and cut the timber into logs of standard length, leaving only a 3-m strip on the shoreward side to prevent erosion of the bank. These logs are ferried to charcoal kilns in a nearby village.

Because the blocks are allocated for clearance in such a way that they are always surrounded by mature forest, repopulation with mangrove propagules occurs rapidly. The debris resulting from the clearing operation takes about 2 years to decompose. After 1 year, the site is inspected. If necessary, natural regeneration is assisted by artificial planting, mainly with *Rhizophora apiculata*. Local villagers are contracted to rear suitable seedlings in small nurseries for this purpose. Weed species can also be removed at this stage. The mangrove fern *Acrostichum* can be a particular problem: it is well adapted to occupying sunlit spaces in the forest, so rapidly latches on to a cleared site, and makes it difficult for mangrove propagules to establish themselves. Destruction of seedlings by crabs and monkeys (pp. 110 and 95) can be a problem. The following year, the site is again inspected, and any parts where seedling survival has been less than 75% successful are again replanted.

Fifteen or so years later, the site is revisited, and the young trees thinned out to a distance of 1.2 m, using a measuring pole of that length. (In pre-metric days the thinning distance was set at the round figure of 4 feet, and there has been no reason to change it). The thinnings—all the same age, hence a standard thickness—are valuable as fishing poles. When the stand is 20 years old, it is again thinned, this time to a distance between trees of 1.8 m (6 feet): this time the thinnings (still of uniform thickness) are of a size suitable for the construction of village houses. Because the previous

Fig. 11.1 A mature stand of mangroves (mainly *Rhizophora*) at the edge of a tidal creek in the Matang, western peninsular Malaysia. The trees are all virtually the same age and size, a result of the management regime. The understorey vegetation consists largely of the mangrove fern *Acrostichum*.

thinning means that the trees are not crowded, these are ideal for their purpose. Finally, after 30 years, the block is again ready for clear felling for charcoal (Figure 11.1; Gan 1993; Ong 1995).

The success of the Matang management lies in the allocation of blocks so that no two adjacent blocks are clear felled within a short time of each other. The entire forest is a mosaic of patches of different ages, apart from the Virgin Jungle Reserve and areas set aside for research.

Since management began, there has been a trend towards virtual monoculture of *R. apiculata* in the intensively managed areas of the Matang. During this time, there is some equivocal evidence of a slight decline in productivity (see p. 203), but overall the Matang is a model of sustainable management of a natural resource—a depressingly rare situation.

In 1992, wood extraction amounted to more than 450 000 t and was worth a little over £2.5 million. In recent years, declining demand for charcoal, and a shortage of workers for the labour-intensive business of timber extraction and charcoal-burning suggest that the future value of the Matang may lie in other products. Although the management is largely directed towards timber extraction, this accounts for only around 12% of the total economic value. The area supports thriving fisheries. The offshore waters annually yield more than 50 000 t of fish, valued at £16.9 million, and supporting nearly 2000 people. Farming of the blood cockle (*Anadara granosa*) currently runs at more than 34 000 t a year, worth £1.8 million, and could be developed further. The Matang also has considerable potential for tourism, being rich in wildlife, including otters, monitor lizards, a wide range of birds including the rare milky stork (*Mycteria cinerea*), and flashing fireflies. At present, with virtually no infrastructure, tourism probably brings in around £250 000 annually to the local economy (Gopinath and Gabriel 1997).

Even if the market for charcoal disappears, managed mangrove forests such as the Matang should therefore have good long-term prospects, provided the connections between mangrove production and other activities are fully recognized.

11.1.3 *Shrimps versus mangroves?*

There are many causes of loss of mangroves. Of these, clearance of mangroves for aquaculture, particularly of shrimps, has caused the most controversy. The issues are complex, and the debate between aquaculturists and conservationists often ferocious. Vehemence is no substitute for argument, however. Without getting too embroiled in controversy, this section will examine the relations between shrimp aquaculture and mangrove depletion. Among the issues are the extent to which mangrove loss can actually be attributed to the spread of aquaculture, the overall environmental impact of aquaculture, and whether sustainable mangrove-based aquaculture has been, or can be, achieved.

Aquaculture is not new. Cultivation of fish and shrimps has been carried out in Asia for many centuries. A good example is shown by the *gei wai* system of southern China. *Gei wai* are shallow ponds formed by impounding areas of mangrove with mud embankments. Deep drainage channels (1–3 m) are dug round the edge of each pond, surrounding a patch of untouched mangroves, and connect to the sea through a sluice. The growth of shrimp (*Penaeus* and related species) is supported by the mangroves in and around the pond. Often fish and molluscs are cultivated simultaneously in the same ponds. To harvest the shrimp, a net is placed across the sluice, and the pond partly drained at a low spring tide. Shrimp concentrate in the drainage channels; large shrimp are collected, and small ones returned to the pond. As the tide rises, the pond floods again, and the net across the sluice excludes predators, while allowing shrimp larvae to enter to restock the pond. Annual production is not high, less than 500 kg/ha, but running costs are low. Shrimp larvae enter with the tide and do not need to be caught at sea or reared in hatcheries, while nutrients for shrimp growth are supplied either by mangrove litterfall or by the incoming tides. Most importantly, the process is sustainable, as the mangrove environment is not greatly altered and mangrove productivity is maintained.

Over the last few decades, aquaculture has burgeoned into a so-called Blue Revolution, in response to a combination of increased international demand, the quest for greater profit, and concern about overfishing of natural stocks. Shrimp farming has increased at an annual rate of 20–30%, and is a global industry with an annual production of more than 700 000 t and retail value of over US$20 million (Primavera 1997b).

Traditional small-scale rearing has given way to more intensive methods of production. The so-called extensive system of rearing is similar to traditional methods, albeit on a larger scale; typical rearing ponds range from 5 to 50 ha. The increased size and intensity of rearing means that ponds cannot be adequately stocked by larvae brought in by the tide. Artificial seeding, usually with larvae caught at sea, is necessary. The food supply is often artificially increased by the addition of fertilizer, and water circulation must be boosted with circulation pumps. Yields increase somewhat, but costs are higher than with traditional methods.

A more fundamental problem is that shrimp depend on mangroves as a nursery area, and as an ultimate productivity base (p. 178). Removal of mangroves to create shrimp ponds undermines the basis of shrimp production. It becomes progressively harder to obtain shrimp larvae to stock the ponds, and the natural food supply brought in by the tides is reduced.

Semi-intensive rearing takes the process further, increasing possible yield by a factor of 10. To achieve this, the stocking rate is greatly increased by using hatchery-reared larvae, water circulation depends heavily on pumping, and artificial feeds are required. The increased intensity also means that various chemical additives and antibiotics must be used, further increasing the costs. Intensive systems further increase densities: again, correspondingly greater inputs of food, chemical treatment, and water pumping, are necessary. Costs can be high, but annual yields may be as great as 20 000 kg/ha.

Much of the increase in extensive shrimp farming, particularly of the extensive systems, has been at the expense of mangroves. The extent of mangrove clearance specifically due to shrimp farming has been hotly disputed. One estimate is that only 7–8% of global mangrove forest loss is due to shrimp aquaculture. Certainly a great deal of mangrove clearance took place before the 1980s boom in shrimp farming got underway. However, even if this figure is correct, it gives a misleading impression of the local impact of clearance for shrimp ponds. In Indonesia, for example, most of the 300 000 ha currently being used to culture shrimps was formerly mangrove forest, and it is planned to increase this figure to more than 1 million ha (Macintosh 1996). In the Philippines, approximately half of the mangrove loss between 1951 and 1988 resulted from the construction of shrimp ponds (Primavera 1997b). Throughout the world, perhaps 800 000 ha of mangrove have been destroyed specifically for shrimp-pond construction.

Mangrove clearance has many secondary effects. In Vietnam, more than two-thirds of the country's mangroves were destroyed by defoliants during the Vietnam War, and development, particularly for shrimp ponds, subsequently continued the process. The environmental consequences have been severe. Coastal erosion has increased. A coastline denuded of mangroves is more vulnerable to storm damage and the inland intrusion of sea water, damaging crops. Fewer shrimp larvae are available to stock aquaculture ponds, and catches of the lucrative mud crab (*Scylla*) have been reduced, as this species also depends on mangroves. When mangrove mud is exposed to the air during pond excavation, it becomes progressively more acid, reducing water quality and shrimp yield. Finally, the increased number of stagnant pools has provided excellent breeding prospects for mosquitoes, and malaria has increased (Macintosh 1996).

Apart from such obviously damaging consequences, the more intensive methods of aquaculture leave an extensive ecological footprint. Although there has been a marked increase in the capacity to produce larvae in hatcheries, many ponds are still stocked with larvae caught at sea. The fine-meshed nets used inevitably trap large quantities of other invertebrates and fish larvae as a by-catch, often amounting to more than 20 times the actual weight of shrimp larvae caught. The impact on other fisheries, and on coastal ecology in general, has not been assessed, but may be considerable.

When pond-reared shrimps are artificially fed, around 30% of the food remains uneaten. The rest—together with a rich brew of chemicals and antibiotics—ends up as effluent. If not excessively concentrated, this waste—rich in nitrogen and phosphorus—represents nutrient which can potentially stimulate mangrove productivity. Depending on the precise method of culture, a 1-ha shrimp pond requires between 2 and 22 ha of neighbouring mangrove forest to assimilate its effluent. Taking into account the other ways in which a mangrove ecosystem can support shrimp-pond functions, an alternative estimate is that each hectare of semi-intensive shrimp farm requires 35–190 ha of adjoining ecosystem, mainly mangrove (Robertson and Phillips 1995; Kautsky *et al.* 1997). If effluent production is greater than can be readily assimilated by surviving mangroves, its ecological impact can be disastrous.

Shrimp farming, then, is not without its problems. In several countries, the Blue Revolution reached Klondyke proportions, economic and ecological problems became apparent, and an unsustainable industry crashed. Taiwan is a good example. In the late 1960s, shrimps became the dominant aquaculture crop. By 1980, around 2000 ha of ponds were producing 5000 t of shrimp. Hatchery production of larvae rose from 6000 in 1968 to 300 million in 1979, to 1.3 billion in 1984 and 3 billion in 1985. By 1987 annual shrimp production reached 100 000 t. By the mid-1990s the industry had virtually collapsed. Overstocking, overcropping, misuse of processed feeds and antibiotics, problems with effluent disposal, and a variety of bacterial and viral diseases caused a fall in production to 30 000 t in 1989. Similar booms followed by slumps have occurred in China, Thailand, and other shrimp-producing countries (Baird and Quarto 1994; Primavera 1997b).

When a shrimp industry collapses, what is left is often a devastated coast, with derelict ponds in place of a previously extensive mangrove habitat. These can sometimes be rehabilitated, but the process takes care, time, and investment resources which are often not available.

The often disastrous history of shrimp farming in mangrove areas does not mean that sustainable cultivation of shrimp is impossible. It is clear that constructing shrimp ponds at the expense of mangrove is not sensible. In fact much of recent shrimp-pond construction has actually been on the landward side of the mangrove belt, above the high-tide mark. If pond area is not too great in relation to the surviving area of mangrove habitat, and if water circulation is appropriately designed, it would even be possible for mangroves to be used as an efficient processing system to limit damaging effects of effluent on the environment. With hatchery-reared larvae to stock the ponds, the destructive trapping of larvae at sea would be largely unnecessary.

Most of the serious problems of shrimp-farming could, in principle, be solved: sustainable aquaculture, without massive destruction of mangroves, is possible.

11.1.4 Mangroves and pollution

Pollution comes in many forms, including exposure to hot-water outflows, toxic heavy metals, pesticides, sewage, or oil spills. Sometimes the contamination is accidental, sometimes deliberate, because mangroves are seen as valueless, fit only for dumping unwanted wastes.

Thermal pollution comes in the form of water from power-plant cooling systems. When *Rhizophora mangle* is exposed to a 5°C rise in water temperature, it responds by decreasing leaf area and increasing the density of aerial roots. Seedlings are more vulnerable than adults to high temperatures (as to low temperatures), and a rise of 7–9°C is sufficient to cause 100% mortality (Lugo and Snedaker 1974; Pernetta 1993). High temperatures also greatly reduce the species richness of the mangrove fauna.

Thermal pollution of mangrove habitats is relatively rare. Various forms of chemical pollution are more frequent, the most serious being heavy metals, pesticides, sewage, and petroleum products. Mangroves often face combinations of these pollutants.

Heavy-metal contamination comes from mine tailings and industrial waste, and includes in particular mercury, lead, cadmium, zinc, and copper. Because assay of heavy metals is simple, there is a great deal of information about their distribution and accumulation within mangrove habitats, and rather less understanding of their biological effects. Much accumulates in mangrove sediments, where it may not be in a position to have profound ecological effects. Mangrove trees themselves may be relatively immune to the toxic effects of heavy metals, but the mangrove fauna may be more vulnerable. Mercury, cadmium, and zinc are acutely toxic to crab larvae and heavy metals in general cause physiological stress and reduced reproductive success, even at sublethal concentrations. Moreover, accumulation in fish, shrimps, or edible molluscs could present serious health problems for the human population (Ellison and Farnsworth 1996; Peters *et al.* 1997).

The same is true of herbicides and pesticides, which usually enter the mangal in run-off from agricultural land. These may also have both acute and chronic effects on both mangrove animals and plants, and can accumulate in the tissues of food species. However, many of the compounds are strongly absorbed on to sediments, and readily degraded in anaerobic soils (Clough *et al.* 1983).

The impact of sewage pollution on mangroves depends very much on the amounts involved. Mangrove growth and productivity may be limited by the available nitrogen and phosphorus, as well as by high salinity (Chapter 2). Sewage is rich in both nutrients, and is of low salinity. Up to a certain level it would therefore be likely to enhance mangrove productivity; indeed, field trials have shown that mangroves can provide a useful method of waste water treatment. High nutrient loads stimulate excessive algal growth and cause deoxygenation of the water by microbial activity. Algae may grow profusely on pneumatophores and aerial roots, impeding gas exchange, or drifts of the sheet-like green alga *Ulva* smother seedlings and may even physically dislodge them from the soil.

The most obvious damage to mangroves, however, comes from oil pollution. Between January 1974 and June 1990 there were, in the tropics, at least 157 major oil spills (greater than 250 000 l) from ships and barges. More than half of these were close enough to the coast to present major threats to coastal ecosystems such as mangroves (Burns *et al.* 1993). In addition, there were numerous small spillages involving smaller volumes, often occurring during loading or discharging of tankers, or through illicit washing out of tanks at sea.

Fuel oil consists of a complex mixture of components which differ in molecular mass and in volatility. Lighter fractions are generally more toxic, but often evaporate before an oil slick reaches the shore or soon afterwards. The residual heavy components often kill mangroves by coating pneumatophores and aerial roots, clogging lenticels and killing roots by asphyxiation. The immediate result is often defoliation of the affected trees; depending on the level and persistence of the pollution, trees either recover or die.

The extent of the damage obviously depends on the amount of oil reaching the shore, which in turn depends on wave and wind conditions. Tidal régime also makes a difference: in the Caribbean, where the normal range is often only 30–50 cm, narrower zones are affected than in parts of the world with more typical 2–3 m tidal ranges. Shore topography and environmental setting also matter: the area of mangroves affected by a given volume of oil will differ between steep and shallow shores, between straight and convoluted coastlines, and between fringing, estuarine, and overwash mangals (Chapter 4).

If a mangrove forest is heavily blanketed with oil, and suffers locally catastrophic effects on fauna and flora, it is not hard to link cause and effect. It is not so easy to evaluate subtler long-term effects on tree health and productivity, or on community structure, without a thorough understanding of the situation before the impact. A further problem in assessing the consequences of oil pollution is that oil spills are unpredictable and there is rarely any baseline environmental information for an adequate comparison before and after the event.

An exception to this is Bahía las Minas, on the Caribbean coast of Panama. This area, fortuitously, includes the Galeta Marine Laboratory, resulting in much background information about the condition of the mangroves before pollution events. The first of these was in December 1968, when the tanker *Witwater* foundered in rough seas, releasing between 2.8 and 3.8 million litres of fuel oil. Because of the time of year, tidal range and rainfall were both relatively high. This combination of circumstances helped to keep channels and larger streams relatively clear of oil. About 49 ha of mangroves were killed, 4% of the mangroves in the bay. By 1979, most of the deforested areas had been repopulated, with the exception of around 3 ha where the sea had succeeded in encroaching on the denuded shore.

The second major spill came in April 1986, when the rupture of a storage tank at a nearby refinery released around 8 million l of crude oil. At first, this was concentrated in an embayment, where it deposited as a layer 5 mm thick. Approximately half of the trees in this area died within 2 months, leaving few survivors in the lower half of the shore. A few days after the spill winds and rain combined to wash an extensive slick out to sea. This was later washed ashore over a wide stretch of coastline. Because of low tidal range at the time, the effects were concentrated in a relatively narrow band bordering the shoreline. Overall, an estimated 69 ha of mangrove trees died. There was also massive mortality of the intertidal fauna, with sessile animals suffering more rapid, as well as long-term, effects than mobile species. Aerial roots of *Rhizophora* were reduced by between one-third and three-quarters, reducing the area available for epifaunal communities (Ellison and Farnsworth 1996).

Acute effects were obvious in the immediate aftermath of the 1986 spill. Some of the deforested area was rehabilitated by artificial replanting, although again a few hectares were irrevocably lost by encroachment of the sea. Several years after the 1986 spill, large areas of low density open canopy remained, suggesting that natural regeneration was slowed by long-term sublethal effects of the oil. One factor delaying regeneration was probably the unusually high mortality of seedlings. Taking this into account, the area affected by the 1986 spill was approximately 377 ha, or 42% of all mangroves in the bay. The total area affected was therefore five to six times greater than that completely cleared (Duke *et al.* 1997).

Some of the oil was absorbed into the sediment, and released intermittently over the following years, so that the initial spill cast a long shadow of

Table 11.2 Generalized response of mangrove forests to oil spills. Compare with Figure 4.4. From Lewis, R.R. (1983). Impact of oil spills on mangrove forests. In Tasks for Vegetation Science, vol. 8 (ed. H.J. Teas) pp. 171–183, W. Junk, with kind permission from Springer Science and Business Media.

Stage	Impact
Acute	
0–15 days	Deaths of birds, fish, and invertebrates
15–30 days	Defoliation and death of small (<1 m) mangroves; loss of aerial-root community
Chronic	
30 days–1 year	Defoliation and death of medium (<3 m) mangroves; tissue damage to aerial roots
1–5 years	Death of larger mangroves; loss of oiled aerial roots; new aerial roots deformed?
1–10 years	Reduced litterfall; reduced reproduction; reduced survival of seedlings; reduced growth or death of recolonizing saplings?; increased insect damage?
10–50 years	Complete recovery of forest?

chronic ecosystem effects. Five years after the 1986 event, secondary oiling was still taking place, and oil levels in bivalve molluscs were still high. A reasonable estimate is that a mangrove forest may take 20–30 years to recover fully from a major oil spill (Table 11.2). As has already been shown in Bahía las Minas, a mangrove forest may be hit by a second oil spill before it has recovered from the first.

11.1.5 Hurricanes and typhoons

Not all destructive impacts on mangroves are caused by humans. Tropical coastlines on the western edge of the major oceans are frequently struck by cyclonic storms. In the Caribbean and Gulf of Mexico these are termed hurricanes; in the western Pacific, typhoons. In either case they are characterized by extremely high wind speeds, torrential rainfall, and storm surges several metres above the normal sea level. In the season, there may be several typhoons or hurricanes a month, with particularly serious ones every few years. Vietnam, for instance, has 8–10 typhoon strikes each year, with winds of more than 100 km/h and tides 2–3 m higher than normal. Heavy rains may sometimes increase sediment movement from the land, accelerate the rate of accretion of a shore, and enable mangrove expansion. In most circumstances, however, the impact of typhoons is highly destructive. The margins of the Indian and Pacific Oceans are also affected by tsunamis, which may also devastate mangroves (see p. 210).

Mangrove forests can absorb much of the energy of the average cyclone, but a severe hurricane can be devastating. Hurricane Andrew passed across the south-western coast of Florida in August 1992, at wind speeds greater than 240 km/h, accompanied by a 5-m storm surge. Heavy damage was caused to about 150 km^2 of mangroves. About 60% of the trees, particularly the larger ones, were either uprooted or broken; of the upright and unbroken trees 25% were dead and 86% defoliated. Many of the surviving trees subsequently died. Initially about 20% of large *Avicennia* trees died, but this rose to 50%, or even higher, over the following year (McCoy *et al.* 1996).

Such a major disturbance is likely to have significant long-term effects on the ecosystem. Of the three major mangrove species, *R. mangle* survived better than *Avicennia germinans*, which in turn fared better than *Laguncularia racemosa*. Selective destruction therefore had a noticeable effect on the species composition of the surviving trees.

The damaged plots were rapidly recolonized by seedlings. In many heavily damaged plots, recolonization was mainly by *Laguncularia*, in others by *Rhizophora*, local conditions dictating which species predominated. Only a few *Avicennia* seedlings appeared. Subsequent seedling mortality was high, and growth slow, particularly in the most disturbed areas. This indicated long-lasting effects of Hurricane Andrew on the mangrove community, with slow recovery and a significant change in community structure. However, hurricanes are not the only events affecting the Florida mangrove community. In January 1997 unusually low temperatures selectively killed smaller *Laguncularia* and *Rhizophora*, while the more cold-tolerant *Avicennia* survived.

A detailed model has been developed of the interactions between the three mangrove species of Florida following hurricane damage. This takes account of recruitment, growth, and survival rates, and how these vary in response to different conditions of salinity, and nutrient and light availability. The model predicted that an open space within the forest canopy would be dominated initially by *Laguncularia* seedlings, but that over time this species would be replaced by others, culminating in dominance by *Avicennia*. This is exactly what happened following Hurricane Andrew (and, previously, Hurricane Donna in 1960). Dominance by a single species is inhibited, apparently, by the intermittent impacts of hurricanes which create gaps in which the successional process can begin again (Chen and Twilley 1998).

Small trees survive hurricanes better than large. Forests in the hurricane zone, such as those of Florida and Puerto Rico, show a fairly uniform tree height, with no exceptionally large trees emerging from the canopy. In Panama—outside the hurricane zone—emergent trees are common, and total forest biomass can be twice that of comparable forests in Florida or Puerto Rico (Lugo and Snedaker 1974).

Intermittent meteorological events are natural experiments on a scale that an ecologist would shrink from contemplating, and reveal much about the interaction between the different tree species and their environment. Small trees survive hurricanes better than large, so that small size and early maturity are promoted. Some species are favoured at the expense of others: *Rhizophora* is more resilient when exposed to high winds, *Laguncularia* is a good colonizer, and *Avicennia* the most cold-tolerant. Such irregular events can therefore be important in shaping community structure.

11.1.6 Mangrove rehabilitation

Although throughout the world mangrove habitats are being drastically reduced in area, there are some exceptions to the trend. In parts of New Zealand and Australia, for instance, *Avicennia* patches are spontaneously expanding. Although the reasons are not clear, the expansion is probably a response to inadvertent alteration of sedimentation and other features of the environment by human activities.

In many countries great efforts are being made to restore previously destroyed mangrove habitats by replanting programmes, or even to plant mangroves where none were known before. The reasons are various. In some cases, the purpose is to conserve or recreate an ecosystem for its own sake. More commonly, replanting is carried out because of an awareness of the value of the mangrove resource for fisheries or other activities, or to curb coastal erosion.

Mangrove species which produce large propagules, such as *Rhizophora*, are good candidates for replanting, since in principle little needs to be done besides removing propagules from one site and inserting them into the mud at another. In practice, matters are rather more complicated. Newly established propagules often suffer heavy mortality from crab or monkey attack. Death can result from toppling by oysters, barnacles, algae, or other organisms settling on the seedlings. These problems are largely overcome by growing the propagules in nurseries for a few months and planting them out when large and robust enough to cope. Sometimes, though, fouling organisms must still be removed by hand.

Choice of a replanting site is critical. Appropriate conditions of soil texture, salinity, and hydrology are essential. If mangroves have disappeared from a certain location because human activity has rendered it untenable, attempts at replanting are unlikely to be successful. Despite these limitations, some tens of thousands of hectares have been successfully replanted, for instance in the USA, the Philippines, Thailand, Panama, Kuwait, and Pakistan (Field 1998; Lewis 2005).

11.1.7 Mangroves of the Indus Delta: a case study

The mangroves of the Indus delta, Pakistan, exemplify most of the issues discussed earlier in this chapter, and make a useful case study (Hogarth 1999).

Without the River Indus, Pakistan would be a desert. The river winds its way southwards through arid plains, to debouch into the Arabian Sea through a complex delta of tidal creeks and inlets. The bulk of Pakistan's mangroves are found in this delta, the largest area of arid zone mangroves in the world. Annual rainfall is negligible, often below 200 mm (much of it falling in a single day!), and less than the evaporation rate. River water is the main source of fresh water, and parts of the delta which do not receive significant river flow become more saline than the sea itself.

Rivers in alluvial plains are generally unstable. Silt builds up, raising the river above the surrounding plain but constraining it within banks of sediment. Periodically, these banks burst, flooding the adjoining plain. The river then establishes a new bed, and the process is repeated. Alluvial deltas are also unstable: channels become blocked with sediment and mud banks are eroded by wave action, so that the principal distributary channels shift around.

In the case of the Indus, geological events have also changed the course of some of the delta channels. The result is that the outflow has varied between more than a dozen separate channels, and the principal outflow has ranged between Karachi and the Indian border (Figure 11.2). A river

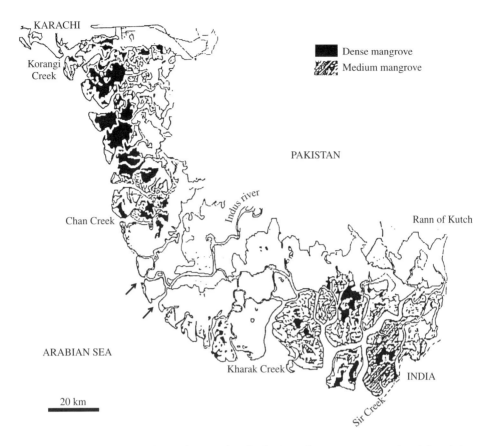

Fig. 11.2 The Indus delta, indicating the distribution of mangrove vegetation inferred from interpretation of satellite imagery. The arrows indicate the current mouths of the Indus. Reproduced with permission of IUCN-Karachi.

that writhes around in this way proved difficult to accommodate within a stable pattern of human settlement. Since the nineteenth century, the Indus has therefore been progressively tamed by the construction of retaining embankments, so that the area now receiving direct river outflow is confined to a so-called active delta of only 119 000 ha.

A further trend has been to divert more and more of the Indus water for irrigation, and for industrial and domestic water supplies. Irrigation engineers measure water volume in a unit known as the MAF, or million acre feet. One MAF is the volume of water that would cover an area of 1 million acres 1 foot deep, equal to 1.23 billion m^3. The MAF, although not metric, at least has the advantage of being on a scale appropriate to the Indus. In the past, annual flow down the Indus has been around 150 MAF; roughly enough water to cover an area the size of the mainland UK more than half a metre deep. The Indus is—or was—a big river.

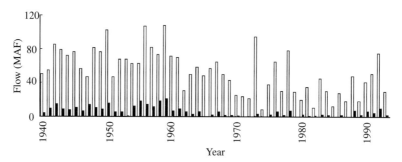

Fig. 11.3 Flow, in MAF (see text for explanation), down the lower Indus from 1940 to 1994. The white and black bars indicate flow during the wet and dry seasons, April–September and October–March, respectively. Data from WAPDA (Pakistan), and Milliman, J.D., Quraishee, G.S., and Beg, M.A.A. 1984. Sediment discharge from the Indus River to the Ocean: past, present and future. In *Marine Geology and Oceanography of Arabian Sea and Coastal Pakistan* (Haq, B.U. and Milliman, J.D., eds), pp. 65–70. Van Nostrand Reinhold, New York.

Since the construction of the latest of the major dams in 1955, the annual flow has sometimes been as low as 30 MAF, and in the dry season virtually no water reaches the coast through the active delta (Figure 11.3). Near Karachi, some water arrives through two small rivers, rich in effluent of various sorts, but at least of low salinity. Elsewhere, virtually the only fresh water arrives from drainage of irrigated land. Reduction of freshwater flow leads to the intrusion of sea water further and further into the delta. The mangroves (and agricultural land within the delta) are exposed to increasing salinity.

Reduced river flow also reduces the amount of sediment delivered, from around 200 million t in the past to less than a quarter of this amount in a typical year. Reduction in sediment delivery affects the equilibrium between deposition and erosion that is a feature of any sandy or muddy coast, causing regression of the shoreline. Reduction in sediment makes it less likely that mangroves will be able to keep pace with rising sea level, which is forecast at 6 mm/year in the Indus delta (Farah and Meynell 1992).

How do these changes affect the mangroves of the Indus delta? To assess this, it is necessary first to have a clear idea of the current status of the mangroves, and how they have changed in the recent past. Satellite imagery is a useful source of information. Different surfaces reflect different wavelengths of light, so that remote sensing techniques can be used to distinguish areas covered by vegetation from those that are bare mud. In principle, it should therefore be a straightforward matter to map the distribution of mangrove vegetation, measure its area, and detect changes with time. In practice, there are serious complications. Different analyses use different methods to identify and categorize mangrove areas. Only rarely is crucial information given about the criteria used, and even more rarely is the analysis 'ground-truthed' by systematically visiting sites and comparing what is visible on the ground with what was inferred from satellite photographs. In some cases algae covering the mud have probably been misinterpreted as mangroves. These, and other complications, explain why estimates of total area sometimes diverge widely.

The best estimate is that the Indus delta currently contains rather less than 200 000 ha of mangrove, and that this area is declining by as much as 2% annually. What satellite images do not reveal is that a significant proportion of the mangal consists of trees stunted by their stressful environment, growing no more than half a metre in height.

In addition to the reduction in area, it appears that diversity has been reduced. Historical records and pollen deposited in the soil suggest that in the past nine species of mangrove occurred within the delta (although there is some doubt about the correct identification of two of these). At present, natural mangal consists almost entirely of *Avicennia marina*, with small and localized patches of only two other species, *Ceriops* and *Aegiceras*.

Is the decline in area and diversity due solely to deprivation of fresh water? This cannot be the whole story. If lack of fresh water was the only factor, mangroves should survive better in the area still served directly by the Indus. This is not the case: the active delta is in fact virtually devoid of mangroves (Figure 11.2). Why should this be? The areas on either side of the active delta are controlled by the Forestry Department of the provincial government. Control is imperfect, but does go some way towards limiting the use of mangroves for firewood and as grazing for domestic animals. This is not the case in the active delta, where virtually no regulation exists. Around 26 000 camels, sheep, goats, water buffalo, and cattle are believed to spend at least part of the year in the area. These feed on *Avicennia* (Figure 5.7), demolishing branches and trampling seedlings, and compact the soil to the extent that it becomes virtually impossible for propagules to establish themselves. Overgrazing, the result of poor management, is a major factor in the destruction of the mangroves.

In other parts of the delta, mangroves have also suffered, but survived. Near Karachi, they have been cleared for harbour and other developments, or been buried in soil from dredging. Overexploitation has opened up cleared areas round many of the villages. Oil, heavy metals, pesticide residues, domestic sewage, animal waste, and assorted industrial effluents enter the creeks around Karachi by hundreds of millions of litres daily. Only a few kilometres from the outfalls, mangrove trees seem to flourish.

What are the consequences of the decline in Indus delta mangroves? At least 100 000 people in the delta depend on mangroves for firewood and fodder, and any further decline in mangroves would affect them adversely. Apart from their exploitation for firewood and fodder, mangroves serve to limit coastal erosion. Pakistan's second largest harbour, Port Qasim, is protected by a mangrove-covered mud bank. Outside this natural harbour defence waves reach a height of 6 m; inside, the maximum height is 0.5 m. Without such coastal defences it is hard to see how Port Qasim could continue to function.

There are also likely to be less-local impacts of mangrove loss. Many thousands of birds overwinter in the delta, having migrated down the Indus valley from Central Asia. Reduction of crucial winter feeding grounds would have ecological effects many thousands of kilometres away.

As in other parts of the world, the Indus mangroves sustain offshore fisheries. The shrimp fishery, mainly for export, is worth around $60 million annually, and other fisheries are probably worth a further $30 million or so. One useful measure of the state of a fishery is the catch per unit effort (CPUE). Figure 11.4 shows the change in CPUE for fish caught before and after completion of the last major dam (so far), in 1955. The shrimp fishery shows a similar pattern. Other factors, such as overfishing, contribute to a declining CPUE, but CPUE fell sharply before expansion of the fishing fleet. Completion of the dam and reduction of river flow are therefore likely to have been significant factors. It is not possible to assess the relative effects of declining mangroves, and of the reduction in fresh water and sediment.

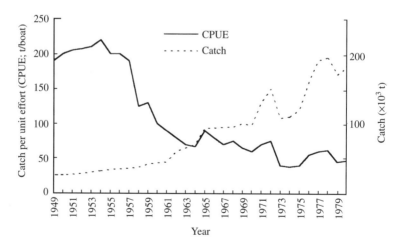

Fig. 11.4 Yearly variations in total fish catch ($\times 10^3$ t), and in catch per unit effort (CPUE; t/boat) along the coast of Sindh between 1949 and 1980. Construction of the Kotri barrage was in 1955. Data from Milliman, J.D., Quraishee, G.S., and Beg, M.A.A. 1984. Sediment discharge from the Indus River to the Ocean: past, present and future. In *Marine Geology and Oceanography of Arabian Sea and Coastal Pakistan* (Haq, B.U. and Milliman, J.D., eds), pp. 65–70. Van Nostrand Reinhold, New York.

Fig. 11.5 Young *Rhizophora apiculata* planted out in a denuded area of the Indus delta near Karachi, Pakistan. To increase tidal inundation frequency, the seedlings have been planted along a shallow irrigation ditch.

Fig. 11.6 Walkway constructed as part of an ecotourism project among the mangroves of the Indus delta, close to Karachi. The mangroves on either side of the walkway are not seedlings, but adult trees, stunted in their growth by environmental factors such as hypersalinity, pollution, and soil compaction.

Efforts are being made to reduce or even reverse the process of decline by a replanting programme. The aims were discussed with local villagers at all stages: without local understanding and involvement, replanting projects have little chance of success. Initially, *Rhizophora mucronata* propagules were collected from elsewhere in Pakistan and grown in nurseries for up to a year before being planted out on selected and carefully prepared sites. During their first years of life, the seedlings are given intensive care, with the removal (by hand) of algae, propping up of any toppled seedlings, and judicious pruning to encourage vigorous growth of a main stem (Figure 11.5). Given this assistance, the survival rate is better than 95%, and after 3 years or so the planted mangroves reproduce for the first time. Although it is too early to evaluate the long-term success of the project, around 12 000 ha have now been replanted (Qureshi 1996). The mangrove-replanting nurseries have been made a feature of a small ecotourism project in the area, to help raise awareness of the significance of mangroves (Figure 11.6).

11.2 Seagrasses: benefits and threats

Compared with mangroves, direct use of seagrass is relatively unimportant. Seeds and rhizomes have been harvested for food, and leaves as fertilizer or animal fodder, or as packing or insulating material. The indirect value of seagrasses to humans is considerably more important. As discussed in earlier chapters, many commercially important species of mollusc, crustacean, and fish depend on seagrass meadows for shelter and nutrition.

Possibly even more important are the functions of seagrass beds in reducing wave and current action. This encourages the settling of sediment, which is trapped and consolidated within the seagrass bed, along with any contaminants. The process locally improves water quality and may protect coral reefs from sedimentation (Green and Short 2003).

Seagrasses are exposed to a number of threats, from both natural processes and human activities. Hurricanes and cyclones, often accompanied by large waves, can cause extensive damage. After one cyclone in 1985 in the western Gulf of Carpentaria (Australia) an estimated 183 km^2 of seagrass was reduced to 33 km^2: a loss of 82%.

Biotic factors can also cause drastic declines in seagrass beds. Overgrazing by sea urchins or molluscs can eliminate quite wide areas of seagrass, possibly when their own natural predators have been reduced by human activity (Hemminga and Duarte 2000).

Zostera marina is also susceptible to a wasting disease caused by a marine slime mould *Labyrinthula zosterae*. In the middle of the last century this reduced *Zostera* coverage by over 90% and virtually eliminated the species over large areas of northern Europe. *Labyrinthula* occurs widely on seagrasses throughout the world, normally causing no harm. It may have become a serious pathogen only because the plants were vulnerable as a result of environmental stresses, possibly human-made (Hemminga and Duarte 2000).

Humans, directly or indirectly, are undoubtedly the major cause of seagrass decline. Dredging shipping channels directly destroys seagrass beds, whereas almost any coastal construction activity results in a high sediment load in coastal waters which can bury smaller seagrass shoots, reduce photosynthesis by obscuring the light, and place seagrasses under considerable physiological stress even if they are not killed outright. Sediment run-off from the land can be increased by overgrazing of domestic animals, while agriculture and human waste can increase nutrient levels in the water, leading to eutrophication and, potentially, catastrophic loss of seagrasses.

Accurate assessment of the extent of seagrass decline is handicapped by the lack of detailed surveys. A widely accepted estimate is that, from the mid-1980s to the mid-1990s alone, up to 12 000 km^2 of seagrass was eliminated worldwide (Short and Willey-Echeverria 1996).

11.3 Global climate change

It is now accepted that the world's climate is changing. There is disagreement—often fierce—over the precise causes, and particularly over the contribution of human activities to climatic change. Whatever the causes, what is the nature of climatic change? The concentration of carbon dioxide in the atmosphere has increased, and mean global temperature and sea level have both risen, and are likely to continue to do so. Other effects, such as changes in the pattern of rainfall, and on the frequency and severity of tropical storms, are less clear cut. How might mangroves and seagrasses be affected by climatic change?

11.3.1 Rise in atmospheric carbon dioxide

Measurements of atmospheric carbon dioxide indicate a rise from around 280 parts per million (ppm) in the late eighteenth century to 368 ppm in 2000. The rise continues. Much is due to human activities, notably the burning of fossil fuels. There are aspirations to reduce emissions, but carbon dioxide levels will continue to increase for the foreseeable future (Watson *et al.* 2001).

Plants depend on carbon dioxide for photosynthesis and might be expected to respond to a rise in atmospheric carbon dioxide levels with increased photosynthesis and growth. This has been demonstrated in some, but not all, mangrove species tested. Carbon dioxide assimilation interacts in complex ways with other aspects of mangrove physiology. The stomata through which carbon dioxide enters a leaf for photosynthesis are also the main route of water loss in transpiration. At higher atmospheric carbon dioxide levels, stomatal conductance is reduced and water loss falls while carbon dioxide uptake levels are maintained. The result is an increase in water-use efficiency (p. 22). The trade-off between water use and carbon dioxide acquisition means that the mangrove response to high atmospheric carbon dioxide may combine increased water-use efficiency with varying effects on transpiration rate and growth, depending on other circumstances.

Elevated carbon dioxide similarly causes greater nitrogen-use efficiency. In general, a rise in atmospheric carbon dioxide can be expected to stimulate growth when it is limited by water, carbon, or nitrogen, but not when salinity is too high for water uptake to be maintained, or when some other nutrient is limiting (Ball and Munns 1992; Field 1995).

Short-term experiments are of limited value in predicting long-term trends. When plants are exposed to elevated carbon dioxide for a long period, the initial increase in photosynthesis is not maintained; compensating mechanisms tend to bring the photosynthetic rate back towards its previous level. Over still longer periods, alterations in plant morphology may also reduce the impact of elevated carbon dioxide. Comparison of herbarium specimens has shown a decrease in the density of stomata (admittedly not in mangrove species) since the Industrial Revolution, which correlates closely with the increase in atmospheric carbon dioxide. With higher prevailing levels of carbon dioxide in the atmosphere, a given level of photosynthesis can be maintained with fewer stomata (Graves and Reavey 1996).

Seagrasses absorb carbon dioxide through the leaf cuticle, rather than by stomata: otherwise, their responses to elevated carbon dioxide levels are likely to be along similar lines, with the possibility of enhanced photosynthesis, and complex interactions with other physiological functions, and with concomitantly changing aspects of the environment, such as rising temperature (Short and Neckles 1999).

To the extent that different species of mangrove and seagrass respond differently, community composition might alter. On balance, however, it seems likely that the direct effects of rising carbon dioxide will have less impact on mangroves and seagrasses (and their fauna) than other features of climate change (Hogarth 2001).

11.3.2 Global warming

Since about the mid-eighteenth century the world has been warming, much of the more recent temperature increase being attributable to the contribution to an enhanced greenhouse effect of elevated levels of carbon dioxide and other so-called greenhouse gases. Human activities have caused atmospheric carbon dioxide to rise (see the previous section). Methane is also rising, by around 0.9% annually. The main source is anaerobic respiration in the guts of domestic animals and in swamps and other waterlogged soils, together with a contribution from mining and oil and gas extraction. The greenhouse effect is further enhanced by relatively small amounts of chlorofluorocarbons (or CFCs), which are particularly potent greenhouse gases.

There has been an increase in global mean temperature of around 0.6°C over the last century. Depending on a number of assumptions, it is predicted that temperature will increase by 0.2–0.5°C per decade during the present century. By the year 2025, global mean temperature would therefore be about 1°C above its present value, and 3.5°C by the end of the twenty-first century (with smaller increases in the tropics). Accompanying this will be a substantially wetter atmosphere and generally (but not universally) increased rainfall. It is possible that tropical storms will be more frequent and more severe (Field 1995; Short and Neckles 1999).

Mangrove species differ in their responses to raised temperature. Photosynthesis, and other variables such as leaf production, usually have clear temperature optima, falling off at lower and higher temperatures. The optimal temperature varies between species and between localities. Among Australian species, most peak between 21 and 28°C. *A. marina*, which is found furthest south, has a peak at 20°C, whereas the more tropically distributed *Xylocarpus* peaks at temperatures above 28°C (Hutchings and Saenger 1987). In some plant species, the optimal temperature for photosynthesis has been shown to alter in individual plants with seasonal temperature changes. It is not known how far mangroves can adapt in this way.

Given the range of temperatures that mangroves experience in their daily lives—which may amount to more than 20°C in temperate species—it seems unlikely that the rises predicted will make a great deal of difference to mangrove productivity. Geographical distribution, however, may be affected. The limits in the latitudinal distribution of mangroves correlate closely with temperature (p. 3). An increase in mean global temperature would, therefore, tend to allow northward and southward extension of the geographical distribution of mangroves. This spread would be limited by topography: within the Indian Ocean, for example, there is no land to the south of South Africa, and no sea north of the Red Sea and Arabian Gulf. On the shores of the Pacific and Atlantic, however, a modest expansion of the total range of mangroves might be expected, probably at the expense of salt-marsh habitats. Because species vary in their temperature tolerance, species would presumably spread to different extents and the combination of mangrove species at different locations might alter.

Seagrasses would probably respond in similar ways to rising temperatures, with some modulation of physiological functions, local shifts in the balance between seagrasses and competing species such as algae, and some alteration in geographical distribution of different species.

Raised temperature would affect other aspects of the ecology of mangrove and seagrass habitats, such as the acceleration of litter decomposition, and effects on the physiology and geographical distribution of components of the mangrove fauna. These effects are not likely to be great and are, in our present state of knowledge, unpredictable.

11.3.3 Sea-level rise

One of the effects of global warming is a rise in overall (eustatic) sea level. This stems largely from thermal expansion of the world's oceans, with a smaller contribution from the melting of glaciers and major continental ice caps. The best current estimates suggest that by the year 2050 mean sea level will have risen by 5–32 cm above the level in 1990, and that by the end of the century the rise will be 9–88 cm. Because the land rises and falls due to tectonic activity, to isostatic readjustment of the Earth's crust following the post-Ice Age redistribution of ice and water, and to other factors, these eustatic changes in sea level translate into greater or smaller local changes in level (Watson *et al.* 2001).

As far as mangroves are concerned, the impact of sea-level rise depends on the environmental setting (Chapter 4). In an estuary, mangrove trees often show a pattern of species zonation up and down the river, dependent largely on a salinity gradient, which is the outcome of the interplay of river and tidal flows. A rise in sea level would have the effect of shifting the zonation pattern in an upriver direction (offset by any increase in local rainfall which might increase the river flow). Upshore/downshore zonation patterns might also move, in a landward direction. Trees on the seaward fringe would be inhibited by extended submergence times, while those on the landward margins might be able to extend, provided suitable habitat was available (Pernetta 1993).

If neither of these situations applies, the outcome may depend on whether the rate of deposition of sediment can keep pace with the rate at which sea level is rising. In the great river deltas, on high oceanic islands with considerable runoff from rainfall, or in situations where large amounts of marine sediments can be trapped, mangroves may avoid losing the race. On small carbonate-based islands, or arid-zone mangroves such as those of the Red Sea, this is unlikely to be the case.

Seagrass communities will be affected in rather similar ways. In some species, the lower limit of distribution is determined by light penetration: a rise in sea level and increased depth of water will mean that these species will decline at their lower depth limits. In estuaries, increased penetration of sea water may mean that seagrasses will move into the estuary to replace salt-intolerant aquatic plants.

Past changes in sea level, and ecological changes associated with them, can give valuable clues to probable future changes. For mangroves—but not for seagrasses—the fossil record is sufficiently continuous for reconstruction of past events. In northern Australia, a dramatic rise in sea level between 8000 and 6000 years ago flooded the coastal plains. In some cases deposition of sediment (mainly originating in the sea) kept pace with the rising sea level and extensive mangrove forests established themselves. When the sea level stabilized, further sediment deposition allowed these to expand seawards (Wolanski and Chappell 1996). Sea-level rises can create opportunities as well as destroy them. A further guide to future prospects is the study of changes known to have occurred in the Caribbean and Pacific regions during the Holocene period. The stratigraphic record of mangroves during this time suggests that mangrove ecosystems could keep pace with a eustatic sea level rise of 8–9 cm/100 years. Between 9 and 12 cm/100 years they are under stress; above this rate of rise the outcome is likely to be complete collapse of the ecosystem (although individual trees could survive). Faced with future rises on the scale predicted, many island mangrove habitats could not persist (Ellison and Stoddart 1991).

Prediction, where many of the key variables to take into account are poorly understood, is a hazardous business. Models of the world's climate and ocean current systems are still not sufficiently sophisticated to give a reliable picture of the immediate future. Forecasts of rainfall, temperature, and sea level a century hence are even less accurate. If the general picture is correct, most mangrove and seagrass communities will be affected by global climate change. Some will adapt with minor adjustments of zonation and species composition, and even some expansion in area. In other areas, species composition will change as geographical ranges expand, with some temperate salt-marsh habitats being replaced by mangroves. In other areas, mangroves and seagrasses might simply disappear (Hogarth 2001).

11.4 What are mangroves and seagrasses worth?

While it is obvious that mangroves and seagrasses are of economic value, putting figures to that value is not straightforward. The total economic value must take account of the whole range of goods and services provided, which differ from one area to the next. Many of these cannot be measured directly (Spaninks and van Beukering 1997; Ewel *et al.* 1998).

Marketable commodities generally can be measured because they have an identifiable volume and cash value. Charcoal production in the Matang mangrove forest is a good example (p. 211). Prices vary with demand and (with exported mangrove products such as shrimp) with exchange rates, but in the short term such fluctuations are generally not a great proportion of the total. Generally these are quoted as gross financial benefits, without taking collecting costs into account.

Products which are not marketed are more difficult to assess: local villagers collect firewood or timber, or catch crabs and fish, for their own use, without money changing hands. These, too, can be estimated, and economic values assigned, often on the basis of the market price that would have been paid for such products had they been purchased.

Problems arise with indirect products whose availability depends on mangroves or seagrasses, but which are harvested at a distance, such as off-shore fisheries. Assigning economic values of mangroves or seagrasses to fisheries means making assumptions about the nature of the linkage between the two: as already discussed (p. 178), the relationship is elusive. In practice, economists usually tacitly accept an established correlation between habitat area and fish yields as implying dependence.

The contribution to aquaculture yields is also problematical. Traditional *gei wai* aquaculture (p. 214) is very largely mangrove-dependent, as is the aquaculture of crabs and molluscs, common in Malaysia, which is based on mesh cages floating in mangrove creeks. Neither technique is destructive of mangroves. Where farmed shrimps are fed artificially, they do not depend on mangroves as an energy source, but may do so as a means of effluent disposal. In this case, too, mangroves supply an essential service in maintaining aquaculture, for which a monetary value can be estimated.

On the other hand, if shrimp ponds are constructed in 'reclaimed' mangrove forest, mangroves can be regarded as a direct alternative to the lucrative cultivation of shrimps. They are therefore assigned a *negative* value, the opportunity cost equivalent to the economic benefit that would accrue from conversion to shrimp ponds, which is being foregone. This looking-glass approach takes no account of the consequences of attempting to realize this value by total removal of mangroves.

Timber, shrimps, and fish are at least tangible products. What of less concrete benefits, such as protection against coastal erosion or incursion by the sea? Values can be assigned, based either on measured erosion after mangrove loss, or on the value of the protected resource such as paddy fields. Should mangroves be credited with the entire value of a commercial port, or of rice production in a coastal plain, because these might be lost if mangroves were eliminated? An alternative approach is based on assessment of the costs of the substitute, such as artificial coastal defences, which would be needed if a given service was no longer provided by mangroves.

Depending on the circumstances, the value placed on coastal protection by mangroves and seagrass beds can be trivial, or astronomical. At one extreme, the loss of the odd hectare may not matter: at the other, loss of protective mangroves could either render a port inoperable, or require massive investment in breakwater construction. An example of this latter situation is discussed on p. 226.

The elements so far discussed are referred to as direct or indirect *use values*: features of mangroves or seagrass beds whose monetary value derives from their use. Direct use values are often relatively straightforward to evaluate, but indirect use values sometimes elusive. Even more intangible are *non-use values*. Examples of non-use values are biodiversity conservation, the ability of mangrove forests to sequester carbon, or the opportunities they afford for research and education. These are generally ignored by environmental economists, but methods have been devised of assigning values to them (although it is not always clear to a mere biologist what these values actually signify).

Some valuations are shown in Table 11.3. The range of figures does reflect genuine variation in the importance of goods and services but also, to a considerable extent, differences in deciding what should be assessed, and how it should be assessed. There is no general agreement on a consistent approach to ecosystem valuation, nor any prospect of precision in measuring many of the variables. Nevertheless, assigning notional monetary values to ecosystem goods and services can be of assistance in clarifying and guiding a decision between two management alternatives within a particular mangrove system, such as whether to develop aquaculture at the expense of preservation. It is of less value in comparing two different ecosystems, where different analytical approaches are employed.

Economic valuation does, however, serve to highlight the importance of mangroves in ways which are generally not appreciated by non-biologists. A recent analysis of the major ecosystems of the world is an example of this (Costanza *et al.* 1997). The value of mangroves and tidal marshes, worldwide,

Table 11.3 Examples of estimates of monetary value of various mangrove systems. Note the range of dates: to be comparable, adjustment should be made for made for variation, with time, in the value of the US dollar. Negative values refer to option use value (see text). Gaps indicate that value was assessed as negligible, or that no attempt was made to assess value. Data from Dixon (1989) and Janssen and Padilla (1996).

	Estimate of monetary value (US$/ha per year)				
	Thailand 1982	Fiji 1976	Indonesia 1992	Philippines 1996	Trinidad 1974
Forestry	30	6	67	151	70
Fisheries	130	100	117	60	125
Agriculture	165	52			
Aquaculture	−2106			−7124	
Erosion			3		
Biodiversity			15		
Local uses	230		33		
Recreation					200
Purification		5820			

Table 11.4 Estimated average value of ecosystem services for mangroves and seagrass habitats. In these estimates, mangroves are combined with salt marsh, and seagrass with algal beds. No useful data are available on the economic value of some possible ecosystem services of mangroves, such as nutrient cycling, climate regulation, genetic resources, or the provision of pollinators, or of many aspects of seagrass beds. From Costanza et al. (1997), with permission of Nature Publishing Group.

Goods/services	Example	Average value (US$/ha per year)	
		Mangrove and salt marsh	Seagrass beds
Disturbance regulation	Protection against coastal erosion, typhoons	1839	
Nutrient cycling			19 002
Waste treatment	Assimilation of effluent from shrimp ponds	6696	
Habitat/refuge	Nursery area for shrimps; habitat for migrating birds	169	
Food production	Energy source for local fisheries	466	
Raw materials	Timber, charcoal	162	2
Recreation	Ecotourism	658	
Total		9990	19 004

was put at just under $10 000/ha per year, and of seagrasses and algal beds (combined) at almost double this value (Table 11.4). The global totals of 1.6×10^{12} for mangroves and 3.8×10^{12} for seagrasses compare with an estimated annual value of the Earth's combined ecosystems of 3.3×10^{13}. The true economic value of mangrove and seagrass ecosystems is enormous.

11.5 Have mangroves and seagrasses a future?

Almost everywhere, mangroves and seagrasses are in retreat, in the face of the relentless pressures of human activity: deliberate or inadvertent destruction, over-exploitation, pollution, and climatic change. Mangroves are particularly vulnerable. If the attrition continues, in many parts of the world mangroves may be reduced to relic patches, too small to support the diversity of organisms characteristic of a thriving mangal. Is there any long-term future for mangrove ecosystems?

Very few mangrove areas are untouched by human activities. The future of mangroves must depend not on remaining isolated from human influences, but on properly regulated interactions with human demands: on effective management. For this to happen, the first step must be education. Unless planners and politicians—as well as local people—appreciate the many contributions of mangroves to human well-being, and the true costs of allowing their destruction, the decline will continue. When mangroves are cleared for timber or for development, the financial gains are rapid but the losses long term. Planning must evaluate all of the costs, as well as the benefits, and look to distant horizons. And those who gain, of course, tend to have more influence on planning decisions than the local people who may lose their means of subsistence.

Some of the pressures on mangroves, such as major water-diversion schemes, or sea-level rise, are not amenable to local management. Here mitigation, rather than prevention, is a realistic objective. Schemes such as the replanting of mangroves in the Indus delta demonstrate how much can be achieved, at relatively low cost. Reproduction and the early settlement stages of mangroves are the most affected by environmental pressures. If these vulnerable stages can be nurtured by artificial methods, a population of adult trees may be able to survive in an environment that is no longer ideal.

Seagrasses, being largely submerged, are less liable than mangroves to overharvesting and deliberate removal, but more liable to inadvertent damage and deterioration in water quality. As with mangroves, survival of seagrass meadows in many parts of the world will depend more and more on adequate understanding of their importance, and sustainable management.

With intelligent and informed management, it is possible to maintain mangroves and seagrasses, not just as a curiosity but as valuable natural resources, managed sustainably so as to continue reaping the benefits without destroying rich and diverse habitats.

Further reading

Perhaps the best introductory book on mangroves for the general reader, with some superb photographs, is

Stafford-Deitsch, J. 1996. *Mangrove. The Forgotten Habitat.* Immel, London.

There are few books that deal thoroughly with mangroves. The following are the most comprehensive. Hutchings and Saenger is specifically about the mangroves of Australia, and Tomlinson, as the title suggests, is a thorough account of mangrove trees, with relatively little on other organisms.

Hutchings, P. and Saenger, P. 1987. *Ecology of Mangroves.* University of Queensland Press, St Lucia.
Tomlinson, P.B. 1986. *The Botany of Mangroves.* Cambridge University Press, Cambridge.

A useful comprehensive review of mangrove ecosystems is

Kathiresan, K. and Bingham, B.L. 2001. Biology of mangrove and mangrove ecosystems. *Advances in Marine Biology* **40**: 85–254.

The world distribution, and many other aspects of mangroves, are comprehensively described in

Spalding, M., Blasco, F., and Field, C. (eds) 1997. *World Mangrove Atlas.* International Society for Mangrove Ecosystems, Okinawa.

Mangroves and seagrasses are discussed in the context of other marine ecosystems in a number of books, of which the following are excellent examples.

Barnes, R.S.K. and Mann, K.H. (eds) 1991. *Fundamentals of Aquatic Ecology.* Blackwell Scientific Publications, Oxford.
Little, C. 1999. *Biology of Soft Shores and Estuaries.* Oxford University Press, Oxford.

For seagrasses, the following are comprehensive and thorough:

Larkum, A.W.D., Orth, R.J., and Duarte, C.M. (eds) 2006. *Seagrasses: Biology, Ecology and Conservation.* Springer-Verlag, Berlin.
Hemminga, M.A. and Duarte, C.M. 2000. *Seagrass Ecology.* Cambridge University Press, Cambridge.

The world distribution and other aspects, are covered in

Green, E.P. and Short, F.T. (eds) 2003. *World Atlas of Seagrasses.* University of California Press, Berkeley.

Websites

The following provide links to a number of informative mangrove and seagrass websites.
www.ncl.ac.uk/tcmweb/tcm/mglinks.htm
wwwscience.murdoch.edu.au/centres/others/mangrove/
www.botany.hawaii.edu/seagrass/
www.ncl.ac.uk/tcmweb/tcm/sglinks.htm

References

Ackerman, J.D. 1995. Convergence of filiform pollen morphologies in seagrasses: functional mechanisms. *Evolutionary Ecology* **9**: 139–153.

Ackerman, J.D. 1998. Is the limited diversity of higher plants in marine systems the result of biophysical limitations for reproduction or evolutionary or physiological constraints? *Functional Ecology* **12**: 979–982.

Alongi, D.M. 1987a. Intertidal zonation and seasonality of meiobenthos. *Marine Biology* **95**: 447–458.

Alongi, D.M. 1987b. The influence of mangrove-derived tannins on intertidal meiobenthos in tropical estuaries. *Oecologia* **71**: 537–540.

Alongi, D.M. 1990. The ecology of tropical soft bottom benthic ecosystems. *Oceanography and Marine Biology: Annual Review* **28**: 381–496.

Alongi, D.M. 2002. Present state and future of the world's mangrove forests. *Environmental Conservation* **29**: 331–349.

Alongi, D.M., Boto, K.G., and Robertson, A.I. 1992. Nitrogen and phosphorus cycles. In *Tropical Mangrove Ecosystems* (Robertson, A.I. and Alongi, D.M., eds), pp. 251–292. American Geophysical Union, Washington DC.

Anderson, C. and Lee, S.Y. 1995. Defoliation of the mangrove *Avicennia marina* in Hong Kong: cause and consequences. *Biotropica* **27**: 218–226.

Baird, I. and Quarto, A. 1994. The environmental and social costs of developing coastal shrimp aquaculture in Asia. In *Trade and Environment in Asia-Pacific, Prospects for Regional Cooperation*, pp. 188–214. Nautilus Institute, Berkeley.

Ball, M.C. 1988a. Ecophysiology of mangroves. *Trees* **2**: 129–142.

Ball, M.C. 1988b. Salinity tolerance in the mangroves *Aegiceras corniculatum* and *Avicennia marina*. I. Water use in relation to growth, carbon partitioning, and salt balance. *Australian Journal of Plant Physiology* **15**: 447–464.

Ball, M.C. and Munns, R. 1992. Plant responses to salinity under elevated atmospheric concentrations of CO_2. *Australian Journal of Botany* **40**: 515–525.

Ball, M.C. and Pidsley, S.M. 1995. Growth responses to salinity in relation to distribution of two mangrove species, *Sonneratia alba* and *S. lanceolata*, in northern Australia. *Functional Ecology* **9**: 77–85.

Ball, M.C., Cowan, I.R., and Farquhar, G.D. 1988. Maintenance of leaf temperature and the optimisation of carbon gain in relation to water loss in a tropical mangrove forest. *Australian Journal of Plant Physiology* **15**: 263–276.

Bandaranayake, W.M. 1998. Traditional and medicinal uses of mangroves. *Mangroves and Salt Marshes* **2**: 133–148.

Banus, M.D. and Kolehmainen, S.E. 1975. Floating, rooting and growth of red mangrove (*Rhizophora mangle* L.) seedlings: effect on expansion of mangroves in southwestern Puerto Rico. In *Proceedings of the International Symposium on*

Biology and Management of Mangroves (Walsh, G.E., Snedaker, S.C., and Teas, H.J., eds), pp. 370–384. University of Florida, Gainesville, FL.

Barlow, B.A. 1966. A revision of the Loranthaceae of Australia and New Zealand. *Australian Journal of Botany* **14**: 421–499.

Barnes, R.S.K. 1967. The osmotic behaviour of a number of grapsoid crabs with respect to their differential penetration of an estuarine system. *Journal of Experimental Biology* **47**: 535–551.

Benner, R., Hodson, R.E., and Kirchman, D. 1988. Bacterial abundance and production on mangrove leaves during initial stages of leaching and biodegradation. *Archiv für Hydrobiologie* **31**: 19–26.

Bertness, M.D. 1985. Fiddler crab regulation of *Spartina alterniflora* production on a New England salt marsh. *Ecology* **66**: 1042–1055.

Bhosale, L.J. and Shinde, L.S. 1983. Significance of cryptovivipary in *Aegiceras corniculatum* (L.) Blanco. In *Tasks for Vegetation Science*, vol. 8 (Teas, H.J., ed.), pp. 123–129. W. Junk, The Hague.

Bingham, B.L. and Young, C.M. 1995. Stochastic events and dynamics of a mangrove root epifaunal community. *Marine Ecology* **16**: 145–163.

Bjorndal, K.A. 1980. Nutrition and grazing behaviour of the Green turtle *Chelonia mydas*. *Marine Biology* **56**: 147–154.

Blasco, F., Saenger, P., and Janodet, E. 1996. Mangroves as indicators of coastal change. *Catena* **27**: 167–178.

Boto, K.G. 1982. Nutrient and organic fluxes in mangroves. In *Mangrove Ecosystems in Australia. Proceedings of the Australian National Mangrove Workshop* (Clough, B.F., ed.), pp. 239–257. Australian Institute of Marine Science, Queensland.

Boto, K.G. 1984. Waterlogged saline soils. In *The Mangrove Ecosystem: Research Methods* (Snedaker, S.C. and Snedaker, J.G., eds), pp. 114–130. UNESCO, Paris.

Boto, K.G. and Wellington, J.T. 1983. Phosphorus and nitrogen nutritional status of a northern Australian mangrove forest. *Marine Ecology Progress Series* **11**: 63–69.

Boto, K.G. and Wellington, J.T. 1984. Soil characteristics and nutrient status in a northern Australian mangrove forest. *Estuaries* **7**: 61–69.

Boto, K.G., Robertson, A.I., and Alongi, D.M. 1991. Mangrove and near-shore connections—a status report from the Australian perspective. In *Proceedings of the Regional Symposium on Living Resources in Coastal Areas* (Alcala, A., ed.), pp. 459–467. University of Philippines, Manila.

Brooks, R.A. and Bell, S.S. 2002. Mangrove response to attack by a root boring isopod: root repair versus architectural modification. *Marine Ecology Progress Series* **231**: 85–90.

Bruce, A.J. 1993. The occurrence of the semi-terrestrial shrimps *Merguia oligodon* (De Man 1888) and *M. rhizophorae* (Rathbun 1900) (Crustacea Decapoda Hippolytidae) in Africa. *Tropical Zoology* **6**: 179–187.

Buck, J. and Buck, E. 1976. Synchronous fireflies. *Scientific American* **234/5**: 74–85.

Bunt, J.S. 1995. Continental scale patterns in mangrove litter fall. *Hydrobiologia* **295**: 135–140.

Burggren, W.W. and McMahon, B.R. (eds) 1988. *Biology of the Land Crabs*. Cambridge University Press, Cambridge.

Burns, K.A., Garrity, S.D., and Levings, S.C. 1993. How many years until mangrove ecosystems recover from catastrophic oil spills? *Marine Pollution Bulletin* **26**: 239–248.

Camilleri, J.C. 1989. Leaf choice by crustaceans in a mangrove forest in Queensland. *Marine Biology* **102**: 453–459.

Cannicci, S., Ritosa, S., Ruwa, R.K., and Vannini, M. 1996a. Tree fidelity and hole fidelity in the tree crab *Sesarma leptosoma* (Decapoda, Grapsidae). *Journal of Experimental Marine Biology and Ecology* **196**: 299–311.

Cannicci, S., Ruwa, R.K., Ritossa, S., and Vannini, M. 1996b. Branch fidelity in the tree crab *Sesarma leptosoma* (Decapoda, Grapsidae). *Journal of Zoology* **238**: 795–801.

Chapman, V.J. 1976. *Mangrove Vegetation*. J. Cramer, Vaduz.

Chen, R. and Twilley, R.R. 1998. A gap dynamic model of mangrove forest development along gradients of soil salinity and nutrient resources. *Journal of Ecology* **86**: 37–51.

Chong, V.C. 1995. The prawn-mangrove connection—fact or fallacy? *Proceedings of Seminar on Sustainable Utilization of Coastal Ecosystem for Agriculture, Forestry and Fisheries in Developing Regions*, pp. 3–20. Malaysian Fisheries Department, Kuala Lumpur

Chong, V.C., Sasekumar, A., Leh, M.U.C., and D'Cruz, R. 1990. The fish and prawn communities of a Malaysian coastal mangrove system, with comparisons to adjacent mud flats and inshore waters. *Estuarine, Coastal and Shelf Science* **31**: 703–722.

Chong, V.C., Sasekumar, A., and Wolanski, E. 1996. The role of mangroves in retaining penaeid prawn larvae in Klang Strait, Malaysia. *Mangroves and Salt Marshes* **1**: 11–22.

Clarke, P.J. 1992. Predispersal mortality and fecundity in the grey mangrove (*Avicennia marina*) in southeastern Australia. *Australian Journal of Ecology* **17**: 161–168.

Clarke, P.J. 1993. Dispersal of grey mangrove (*Avicennia marina*) propagules in southeastern Australia. *Aquatic Botany* **45**: 195–204.

Clarke, P.J. and Myerscough, P.J. 1993. The intertidal distribution of the gray mangrove (*Avicennia marina*) in southeastern Australia—the effects of physical conditions, interspecific competition, and predation on propagule establishment and survival. *Australian Journal of Ecology* **18**: 307–315.

Clarke, P.J., Kerrigan, R.A., and Westphal, J.C. 2001. Dispersal potential and early growth in 14 tropical mangroves: do early life history traits correlate with patterns of adult distribution? *Journal of Ecology* **89**: 648–659.

Clay, R.E. and Andersen, A.N. 1996. Ant fauna of a mangrove community in the Australian seasonal tropics, with particular reference to zonation. *Australian Journal of Zoology* **44**: 521–533.

Clayton, D.A. 1993. Mudskippers. *Oceanography and Marine Biology Annual Review* **31**: 507–577.

Clough, B.F. 1992. Primary productivity and growth of mangrove forests. *Tropical Mangrove Ecosystems* (Robertson, A.I. and Alongi, D.M., eds), pp. 225–249. American Geophysical Union, Washington DC.

Clough, B.F. and Scott, K. 1989. Allometric relationships for estimating aboveground biomass in six mangrove species. *Forest Ecology and Management* **27**: 117–127.

Clough, B.F., Boto, K.G., and Attiwill, P.M. 1983. Mangroves and sewage: a re-evaluation. In *Tasks for Vegetation Science*, vol. 8 (Teas, H.J., ed.), pp. 151–161. W. Junk, The Hague.

Cook, L.M. 1986. Site selection in a polymorphic mangrove snail. *Biological Journal of the Linnean Society* **29**: 101–113.

Cook, L.M. and Freeman, P.M. 1986. Heating properties of morphs of the mangrove snail, *Littoraria pallescens*. *Biological Journal of the Linnean Society* **29**: 295–300.

Cook, L.M., Currey, J.D., and Sarsam, V.H. 1985. Differences in morphology in relation to microhabitat in littorinid species from a mangrove in Papua New Guinea. *Journal of Zoology* **206**: 297–310.

Costanza, R., d'Arge, R., de Groot, R., Farber, S., Grasso, M., Hannon, B. *et al.* 1997. The value of the world's ecosystem services and natural capital. *Nature* **387**: 253–260.

Coupland, G.T., Paling, E.I., and McGuinness, K.A. 2006. Floral abortion and pollination in four species of tropical mangroves from northern Australia. *Aquatic Botany* **84**: 151–157.

Curran, M. 1985. Gas movements in the roots of *Avicennia marina* (Forsk.) Vierh. *Australian Journal of Plant Physiology* **12**: 97–108.

Danielsen, F., Sørensen, M.K., Olwig, M.F., Selvam, V., Parish, F., Burgess, N.D. *et al.* 2005. The Asian tsunami: a protective role for coastal vegetation. *Science* **310**: 643.

Davis, B.C. and Fourqurean, J.W. 2001. Competition between the tropical alga, *Halimeda incrassata*, and the seagrass, *Thalassia testudinum*. *Aquatic Botany* **71**: 217–232.

Dawson, S.P. and Dennison, W.C. 1996. Effects of ultraviolet and photosynthetically active radiation on five seagrass species. *Marine Biology* **125**: 629–638.

Dierenfeld, E.S., Koontz, F.W., and Goldstein, R.S. 1992. Feed intake, digestion and passage of the Proboscis monkey (*Nasalis larvatus*) in captivity. *Primates* **33**: 399–405.

Dittel, A.I., Epifanio, C.E., and Lizano, O. 1991. Flux of larvae in a mangrove creek in the Gulf of Nicoya, Costa Rica. *Estuarine, Coastal and Shelf Science* **32**: 129–140.

Dixon, J.A. 1989. Valuation of mangroves. *Tropical Coastal Area Management* **4**: 1–6.

Dorenbosch, M., Grol, M.G.G., Christianen, M.J.A., Nagelkerken, I., and van der Velde, G. 2005. Indo-Pacific seagrass beds and mangroves contribute to fish density and diversity on adjacent coral reefs. *Marine Ecology Progress Series* **302**: 63–76.

Duarte, C.M. 1991. Seagrass depth limits. *Aquatic Botany* **40**: 363–377.

Duarte, C.M. 1995. Submerged aquatic vegetation in relation to different nutrient regimes. *Ophelia* **41**: 87–112.

Duarte, C.M. 2001. Seagrasses. In *Encyclopaedia of Biodiversity*, vol. 5 (Levin, S.A., ed.), pp. 255–268. Academic Press, San Diego.

Duarte, C.M. 2002. The future of seagrass meadows. *Environmental Conservation* **29**: 192–206.

Duarte, C.M. and Cebrián, J. 1996. The fate of marine autotrophic production. *Limnology and Oceanography* **41**: 1758–1766.

Duarte, C.M. and Chiscano, C.L. 1999. Seagrass biomass and production: a reassessment. *Aquatic Botany* **65**: 159–174.

Duffy, J.E. 2006. Biodiversity and the functioning of seagrass ecosystems. *Marine Ecology Progress Series* **311**: 233–250.

Duke, N.C. 1992. Mangrove floristics and biogeography. In *Tropical Mangrove Ecosystems* (Robertson, A.I. and Alongi, D.M., eds), pp. 63–100. American Geophysical Union, Washington DC.

Duke, N.C. 1995. Genetic diversity, distributional barriers and rafting continents—more thoughts on the evolution of mangroves. *Hydrobiologia* **295**: 167–181.

Duke, N.C., Pinzón, Z.S., and Prada, M.C. 1997. Large-scale damage to mangrove forests following two large oil spills in Panama. *Biotropica* **29**: 2–14.

Duke, N.C., Lo, E.Y.Y., and Sun, M. 2002. Global distribution and genetic discontinuities of mangroves—emerging patterns in the evolution of *Rhizophora*. *Trees* **16**: 65–79.

Dunson, W.A. 1970. Some aspects of electrolyte and water balance in three estuarine reptiles, the diamond back terrapin, American and "salt water" crocodiles. *Comparative Biochemistry and Physiology* **32**: 161–174.

Dunson, W.A. 1978. Role of the skin in sodium and water exchange of aquatic snakes placed in sea water. *American Journal of Physiology Regulatory, Integrative and Comparative Physiology* **235**, R151–R159.

Dye, A.H. and Lasiak, T.A. 1986. Microbenthos, meiobenthos and fiddler crabs: trophic interactions in a tropical mangrove sediment. *Marine Ecology Progress Series* **32**: 259–264.

Dye, A.H. and Lasiak, T.A. 1987. Assimilation efficiencies of fiddler crabs and deposit-feeding gastropods from tropical mangrove sediments. *Comparative Biochemistry and Physiology* **87A**: 341–344.

Edney, E.B. 1961. The water and heat relationships of fiddler crabs (*Uca* spp.). *Transactions of the Royal Society of South Africa* **36**: 71–91.

Elliott, A.B. and Karunakaran, L. 1974. Diet of *Rana cancrivora* in fresh water and brackish water environments. *Journal of Zoology* **174**: 203–215.

Ellison, A.M. 2002. Macroecology of mangroves: large-scale patterns and processes in tropical coastal forests. *Trees* **16**: 181–194.

Ellison, A.M. and Farnsworth, E.J. 1996. Anthropogenic disturbance of Caribbean mangrove ecosystems: past impacts, present trends, and future predictions. *Biotropica* **28**: 549–565.

Ellison, A.M., Farnsworth, E.J., and Twilley, R.R. 1996. Facultative mutualism between red mangroves and root-fouling sponges in Belizean mangal. *Ecology* **77**: 2431–2444.

Ellison, A.M., Farnsworth, E.J., and Merkt, R.E. 1999. Origins of mangrove ecosystems and the mangrove biodiversity anomaly. *Global Ecology and Biogeography* **8**: 95–115.

Ellison, J.C. and Stoddart, D.R. 1991. Mangrove ecosystem collapse during predicted sea-level rise: Holocene analogues and implications. *Journal of Coastal Research* **7**: 151–165.

Elmqvist, T. and Cox, P.A. 1996. The evolution of vivipary in flowering plants. *Oikos* **77**: 3–9.

English, S., Wilkinson, L., and Baker, V. (eds) 1997. *Survey Manual for Tropical Marine Resources*, 2nd edn. Australian Institute of Marine Science, Townsville, Queensland.

Enriquez, S., Marbà, N., and Duarte, C.M. 2001. Effects of *Thalassia testudinum* on sediment redox. *Marine Ecology Progress Series* **219**: 149–158.

Epifanio, C. 1988. Transport of crab larvae between estuaries and the continental shelf. In *Lecture Notes on Coastal and Estuarine Systems*, vol. 22 (Jansson, B.-O., ed.), pp. 291–305. Springer-Verlag, Berlin.

Eshky, A.A., Atkinson, R.J.A., and Taylor, A.C. 1995. Physiological ecology of crabs from Saudi Arabian mangrove. *Marine Ecology Progress Series* **126**: 83–95.

Ewel, K.C., Twilley, R.R., and Ong, J.E. 1998. Different kinds of mangrove forests provide different goods and services. *Global Ecology and Biogeography Letters* **7**: 83–94.

FAO. 1994. *Mangrove Forest Management Guidelines.* Food and Agricultural Organization, Rome.

Farah, A. and Meynell, P.J. 1992. Sea level rise—possible impacts on the Indus Delta, Pakistan. Korangi Ecosystem Project, issues paper 2, pp. 1–20. International Union for the Conservation of Nature, Pakistan.

Farnsworth, E. 2000. The ecology and physiology of viviparous and recalcitrant seeds. *Annual Reviews of Ecology and Systematics* **31**: 107–138.

Farnsworth, E.J. and Ellison, A.M. 1991. Patterns of herbivory in Belizean mangrove swamps. *Biotropica* **13**: 555–567.

Farnsworth, E.J. and Ellison, A.M. 1996. Scale-dependent spatial and temporal variability in biogeography of mangrove root epibiont communities. *Ecological Monographs* **66**: 45–66.

Farnsworth, E.J. and Farrant, J.M. 1998. Reductions in abcsisic acid are linked with viviparous reproduction in mangroves. *American Journal of Botany* **85**: 760–769.

Felgenhauer, B.E. and Abele, L.G. 1983. Branchial water movement in the grapsid crab *Sesarma reticulatum* (Say). *Journal of Crustacean Biology* **3**: 187–195.

Feller, I.C. 1995. Effects of nutrient enrichment on growth and herbivory of dwarf red mangrove (*Rhizophora mangle*). *Ecological Monograph* **65**: 477–505.

Feller, I.C. 1996. Effects of nutrient enhancement on leaf anatomy of dwarf *Rhizophora mangle* L. (red mangrove). *Biotropica* **28**: 13–22.

Fenchel, T., Esteban, G.F., and Finlay, B.J. 1997. Local versus global diversity of microorganisms: cryptic diversity of ciliated protozoa. *Oikos* **80**: 220–225.

Field, C.D. 1984. Movement of ions and water into the xylem sap of tropical mangroves. In *Physiology and Management of Mangroves* (Teas, H.J., ed.), pp. 49–52. W. Junk, The Hague.

Field, C.D. 1995. Impact of expected climate change on mangroves. *Hydrobiologia* **295**: 75–81.

Field, C.D. 1998. Rehabilitation of mangrove ecosystems: an overview. *Marine Pollution Bulletin* **37**: 383–392.

Fleming, M., Lin, G., and Sternberg, L.da S.L. 1990. Influence of mangrove detritus in an estuarine ecosystem. *Bulletin of Marine Science* **47**: 663–669.

Ford, J. 1982. Origin, evolution and speciation of birds specialized to mangroves in Australia. *Emu* **82**: 12–23.

Fourqurean, J.W., Powell, G.V.N., Kenwirthy, W.J., and Zieman, J.C. 1995. The effects of long-term manipulation of nutrient supply on competition between the seagrasses *Thalassia testudinum* and *Halodule wrightii* in Florida Bay. *Oikos* **72**: 349–358.

Fox, A.D. 1996. *Zostera* exploitation by Brent geese and wigeon on the Exe estuary, southern England. *Bird Study* **43**: 257–268.

France, K.E. and Duffy, J.E. 2006. Diversity and dispersal interactively affect predictability of ecosystem function. *Nature* **441**: 1139–1143.

Franklin, C.E. and Grigg, G.C. 1993. Increased vascularity of the lingual salt-glands of the estuarine crocodile, *Crocodylus porosus*, kept in hyperosmotic salinity. *Journal of Morphology* **218**: 143–151.

Fratini, S., Vigiani, V., Vannini, M., and Cannicci, S. 2004. *Terebralia palustris* (Gastopoda; Potamididae) in a Kenyan mangal: size structure, distribution and impact on the consumption of leaf litter. *Marine Biology* **144**: 1173–1182.

Fratini, S., Vannini, M., Cannicci, S., and Schubart, C.D. 2005. Tree-climbing crabs: a case of convergent evolution. *Evolutionary Ecology Research* **7**: 219–233.

Furukawa, K., Wolanski, E., and Mueller, H. 1997. Currents and sediment transport in mangrove forests. *Estuarine, Coastal and Shelf Science* **44**: 301–310.

Gambi, M.C., van Tussenbroek, B.I., and Brearley, A. 2003. Mesofaunal borers in seagrasses: world-wide occurrence and a new record of boring polychaetes in the Mexican Caribbean. *Aquatic Botany* **76**: 65–77.

Gan, B.K. 1993. Forest management in Matang. In *Living Coastal Resources. Proceedings of Workshop on Mangrove Fisheries and Connections* (Sasekumar, A., ed.), pp. 15–26. Australian International Development Assistance Bureau, Canberra.

Gee, J.M. and Somerfield, P.J. 1997. Do mangrove diversity and leaf litter decay promote meiofaunal diversity? *Journal of Experimental Marine Biology and Ecology* **218**: 13–33.

Giddins, R.L., Lucas, J.S., Neilson, M.J., and Richards, G.N. 1986. Feeding ecology of the mangrove crab *Neosarmatium smithi* (Crustacea: Decapoda: Sesarmidae). *Marine Ecology Progress Series* **33**: 147–155.

Gill, A.M. and Tomlinson, P.B. 1977. Studies on the growth of red mangrove (*Rhizophora mangle* L.). 4. The adult root system. *Biotropica* **9**: 145–155.

Gomez, M.A. and Winkler, S. 1991. Bromeliads from mangroves on the Pacific coast of Guatemala. *Revista de Biologia Tropical* **39**: 207–214.

Gong, W.K. and Ong, J.E. 1990. Plant biomass and nutrient flux in a managed mangrove forest in Malaysia. *Estuarine, Coastal and Shelf Science* **31**: 519–530.

Gong, W.K. and Ong, J.E. 1995. The use of demographic studies in mangrove silviculture. *Hydrobiologia* **295**: 255–261.

González-Farias, F. and Mee, L.D. 1988. Effect of mangrove humic-like substances on biodegradation rate of detritus. *Journal of Experimental Marine Biology and Ecology* **119**: 1–13.

Gopinath, N. and Gabriel, P. 1997. Management of living resources in the Matang Mangrove Reserve, Perak, Malaysia. *Intercoast Network* **March** 1997: 23–24.

Gordon, M.S. and Tucker, V.A. 1965. Osmotic regulation in the tadpoles of the crab-eating frog (*Rana cancrivora*). *Journal of Experimental Biology* **42**: 437–445.

Gordon, M.S., Ng, W.W.S., and Yip, A.Y.W. 1978. Aspects of the physiology of terrestrial life in amphibious fishes. III. The Chinese mudskipper *Periophthalmus cantonensis*. *Journal of Experimental Biology* **72**: 57–75.

Gordon, M.S., Schmidt-Nielsen, K., and Kelly, H.M. 1961. Osmotic regulation in the crab-eating frog (*Rana cancrivora*). *Journal of Experimental Biology* **38**: 659–678.

Graves, J.D. and Reavey, D. 1996. *Global Environmental Change: Plants, Animals and Communities*. Longman, Harlow.

Gray, I.E. 1957. A comparative study of the gill area of crabs. *Biological Bulletin* **112**: 34–42.

Green, E.P. and Short, F.T. 2003. *World Atlas of Seagrasses*. University of California Press, Berkeley.

Gunasekar, M. 1993. Changes in chlorophyll content, photosynthetic rate and saccharides level in developing *Rhizophora* hypocotyls. *Photosynthetica* **29**: 635–638.

Hall, L.M., Hanisak, M.D., and Virnstein, R.W. 2006. Fragments of the seagrasses *Halodule wrightii* and *Halophila johnsonii* as potential recruits in Indian River Lagoon, Florida. *Marine Ecology Progress Series* **310**: 109–117.

Hamilton, L.S. and Murphy, D.H. 1988. Use and management of Nipa Palm (*Nypa fruticans*, Arecaceae): a review. *Economic Botany* **42**: 206–213.

Heck, K.L. and Valentine, J.F. 1995. Sea urchin herbivory: evidence for long-lasting effects in subtropical seagrass meadows. *Journal of Experimental Marine Biology and Ecology* **189**: 205–217.

Heck, K.L. and Valentine, J.F. 2006. Plant-herbivore interactions in seagrass meadows. *Journal of Experimental Marine Biology and Ecology* **330**: 420–436.

Hemminga, M.A. and Duarte, C.M. 2000. *Seagrass Ecology*. Cambridge University Press, Cambridge.

Hemminga, M.A., Slim, F.J., Kazungu, J., Ganssen, G.M., Nieuwenhuize, J., and Kruyt, N.M. 1994. Carbon outwelling from a mangrove forest with adjacent seagrass beds and coral reefs (Gazi Bay, Kenya). *Marine Ecology Progress Series* **106**: 291–304.

Higgins, R.P. and Thiel. H. (eds) 1988. *Introduction to the Study of Meiofauna*. Smithsonian Institution, Washington DC.

Hogarth, P.J. 1999. The decline in Indus Delta mangroves: causes, consequences, and cures. In *Mangrove Ecosystems: Natural Distribution, Biology and Management* (Bhat, N., Taha, F.K., and Al-Nasser, A.Y., eds), pp. 65–81. Kuwait Institute for Scientific Research, Kuwait.

Hogarth, P.J. 2001. Mangroves and global climate change. *Ocean Yearbook* **15**: 331–349.

Houbrick, R.S. 1991. Systematic review and functional morphology of the mangrove snails *Terebralia* and *Telescopium* (Potamididae, Prosobranchia). *Malacologia* **33**: 289–338.

Hovenden, M.J. and Allaway, W.G. 1994. Horizontal structures on pneumatophores of *Avicennia marina* (Forsk.) Vierh.—a new site of oxygen conductance. *Annals of Botany* **73**: 377–388

Hughes, A.R. and Stachowicz, J.J. 2004. Genetic diversity enhances the resistance of a seagrass ecosystem to disturbance. *Proceedings of the National Academy of Sciences USA* **101**: 8998–9002.

Hutchings, P. and Saenger, P. 1987. *Ecology of Mangroves*. University of Queensland Press, St Lucia.

Huxley, C.R. 1978. The ant-plants *Myrmecodia* and *Hydnophytum* (Rubiaceae) and the relationships between their morphology, ant occupants, physiology and ecology. *New Phytologist* **80**: 231–268.

Hyde, K.D. and Lee, S.Y. 1995. Ecology of mangrove fungi and their role in nutrient cycling: what gaps occur in our knowledge? *Hydrobiologia* **295**: 107–118.

Icely, J.D. and Jones, D.A. 1978. Factors affecting the distribution of the genus *Uca* (Crustacea: Ocypodidae) on an East African shore. *Estuarine and Coastal Marine Science* **6**: 315–325.

Irlandi, E.A. and Peterson, C.H. 1991. Modification of animal habitat by large plants: mechanisms by which seagrasses influence clam growth. *Oecologia* **87**: 307–318.

Ishimatsu, A., Khoo, K.H., and Takita, T. 1998. Deposition of air in burrows of tropical mudskippers as an adaptation to the hypoxic mudflat environment. *Science Progress* **81**: 289–297.

Jackson, K., Butler, D.G., and Brooks, D.R. 1996. Habitat and phylogeny influence salinity discrimination on crocodilians—implications for osmoregulatory physiology and historical biogeography. *Biological Journal of the Linnean Society* **58**: 371–383.

Jacobs, R.P.W.M., den Hartog, C., Braster, B.F., and Carriere, F.C. 1981. Grazing of the seagrass *Zostera noltii* by birds at Terschelling (Dutch Wadden Sea). *Aquatic Botany* **10**: 241–259.

Janssen, R. and Padilla, J.E. 1996. *Valuation and Evaluation of Management Alternatives for the Pagbilao Mangrove Forest*. CREED Working Paper no. 9, pp. 1–47. International Institute for Environment and Development, Amsterdam.

Janzen, D.H. 1974. Epiphytic myrmecophytes in Sarawak: mutualism through the feeding of plants by ants. *Biotropica* **6**: 237–259.

Janzen, D.H. 1985. Mangroves: where's the understory? *Journal of Tropical Ecology* **1**: 89–92.

Jernakoff, P. and Nielsen, J. 1997. The relative importance of amphipod and gastropod grazers in *Posidonia sinuosa* meadows. *Aquatic Botany* **56**: 183–202.

Jones, C.G., Lawton, J.H., and Shachak, M. 1994. Organisms as ecosystem engineers. *Oikos* **69**: 373–386.

Jones, D.A. 1984. Crabs of the mangal ecosystem. In *Hydrobiology of the Mangal* (Por, F.D. and Dor, I., eds), pp. 89–109. W. Junk, The Hague.

Joshi, G.V., Pimplaskar, M., and Bhosale, L.J. 1972. Physiological studies in germination of mangroves. *Botanica Marina* **15**: 91–95.

Katz, L.C. 1980. Effects of burrowing by the fiddler crab, *Uca pugnax* (Smith). *Estuarine and Coastal Marine Science* **11**: 233–237.

Kautsky, N., Berg, H., Folke, C., Larsson, J., and Troell, M. 1997. Ecological footprint for assessment of resource use and development limitations in shrimp and tilapia aquaculture. *Aquaculture Research* **28**: 753–766.

Kendrick, G.A., Marbà, N., and Duarte, C.M. 2005. Modelling formation of complex topography by the seagrass *Posidonia oceanica*. *Estuarine Coastal and Shelf Science* **65**: 717–725.

Klekowski, E.J., Lowenfeld, R., and Hepler, P.K. 1994. Mangrove genetics. II. Outcrossing and lower spontaneous mutation rates in Puerto Rican *Rhizophora*. *International Journal of Plant Science* **155**: 373–381.

Klumpp, D.W., Salita-Espinosa, J.T., and Fortes, M.D. 1993. Feeding ecology and trophic role of sea urchins in a tropical seagrass community. *Aquatic Botany* **45**: 205–229.

Kuo, J. and McComb, A.J. 1989. Seagrass taxonomy, structure and development. In *Biology of Seagrasses. A Treatise on the Biology of Seagrasses with Special Reference to the Australian Region* (Larkum, A.W.D., McComb, A.J., and Shepherd, S.A., eds), pp. 6–73. Elsevier, Amsterdam.

Kwok, P.W. and Lee, S.Y. 1995. The growth performances of two mangrove crabs, *Chiromanthes bidens* and *Parasesarma plicata* under different leaf litter diets. *Hydrobiologia* **295**: 141–148.

Lacap, C.D.A., Vermaat, J.E., Rollon, R.N., and Nacorda, H.M. 2002. Propagule dispersal of the SE Asian seagrasses *Enhalus acoroides* and *Thalassia hemprichii*. *Marine Ecology Progress Series* **235**: 75–80.

Lanyon, J.M., Limpus, C.J., and Marsh, H. 1989. Dugongs and turtles: grazers in the seagrass system. In *Biology of Seagrasses. A Treatise on the Biology of Seagrasses with Special Reference to the Australian Region* (Larkum, A.W.D., McComb, A.J., and Shepherd, S.A., eds), pp. 610–734. Elsevier, Amsterdam.

Larkum, A.W.D. and den Hartog, C. 1989. Evolution and biogeography of seagrasses. In *Biology of Seagrasses. A Treatise on the Biology of Seagrasses with Special Reference to the Australian Region* (Larkum, A.W.D., McComb, A.J., and Shepherd, S.A., eds), pp. 112–156. Elsevier, Amsterdam.

Lee, S.Y. 1991. Herbivory as an ecological process in a *Kandelia candel* (Rhizophoraceae) mangal in Hong Kong. *Journal of Tropical Ecology* **7**: 337–348.

Lee, S.Y. 1993. Leaf choice of sesarmine crabs, *Chiromanthes bidens* and *C. maipoensis*, in a Hong Kong mangal. In *The Marine Biology of the South China Sea* (Morton, B., ed.). Hong Kong University Press, Hong Kong.

Lee, S.Y. 1997. Potential trophic importance of the faecal material of the mangrove sesarmine crab *Sesarma messa*. *Marine Ecology Progress Series* **159**: 275–284.

Lee, S.Y. 2004. Relationship between mangrove abundance and tropical prawn production: a re-evaluation. *Marine Biology* **145**: 943–949.

Lefebvre, G. and Poulin, B. 1997. Bird communities in Panamanian black mangroves: potential effects of physical and biotic factors. *Journal of Tropical Ecology* **13**: 97–113.

Leh, C.M.U. and Sasekumar, A. 1985. The food of sesarmid crabs in Malaysian mangrove forests. *Malayan Nature Journal* **39**: 135–145.

Les, D.H., Cleland, M.A., and Waycott, M. 1997. Phylogenetic studies in Alismatidae, II: Evolution of marine angiosperms (Seagrasses) and hydrophily. *Systematic Botany* **22**: 443–463.

Lewis, R.R. 1983. Impact of oil spills on mangrove forests. In *Tasks for Vegetation Science*, vol. 8 (Teas, H.J., ed.), pp. 171–183. W. Junk, The Hague.

Lewis, R.R. 2005. Ecological engineering for successful management and restoration of mangrove forests. *Ecological Engineering* **24**: 403–418.

Lin, G.H. and Sternberg, L.D.L. 1994. Utilization of surface water by Red Mangrove (*Rhizophora mangle* L.)—an isotopic study. *Bulletin of Marine Science* **54**: 94–102.

Lin, G.H., Banks, T., and Sternberg, L.da S.L. 1991. Variations in $\delta^{13}C$ values for the seagrass *Thalassia testudinum* and its relations to mangrove carbon. *Aquatic Botany* **40**: 333–341.

Lirman, D. and Cropper, W.P. 2003. The influence of salinity on seagrass growth, survivorship, and distribution within Biscayne Bay, Florida: field, experimental, and modelling studies. *Estuaries* **26**: 131–141.

Lötschert, W., and Liemann, F. 1967. Die Salzspeicherung im Keimling von *Rhizophora mangle* L. während der Entwicklung auf der Mutterpflanze. *Planta* **77**: 142–156.

Lugo, A.E. 1980. Mangrove ecosystems: successional or steady state? *Biotropica* **12s**, 65–72.

Lugo, A.E. and Snedaker, S.C. 1974. The ecology of mangroves. *Annual Reviews of Ecology and Systematics* **5**: 39–64.

Lugo, A.E., Sell, M., and Snedaker, S.C. 1976. Mangrove ecosystem analysis. In *Systems Analysis and Simulation in Ecology* (Patten, C., ed.), pp. 113–145. Academic Press, New York.

Macintosh, D.J. 1982. Ecological comparisons of mangrove swamp and salt marsh fiddler crabs. In *Wetlands, Ecology and Management* (Gopal, B., Turner, R.E., Wetzer, R.G., and Waigham, D.F., eds), pp. 243–257. International Science Publications, Jaipur.

Macintosh, D.J. 1984. Ecology and productivity of Malaysian mangrove crab populations. *Proceedings of the Asian Symposium on the Mangrove Environment—Research and Management 1984*: 354–377.

Macintosh, D.J. 1988. The ecology and physiology of decapods of mangrove swamps. *Symposium of the Zoological Society of London* **59**: 315–341.

Macintosh, D.J. 1996. Mangroves and coastal aquaculture: doing something positive for the environment. *Aquaculture Asia* **October–December**: 3–8.

Macnae, W. 1968. A general account of the fauna and flora of the mangrove swamps and forests in the Indo-Pacific Region. *Advances in Marine Biology* **6**: 73–270.

Mann, K.H. 1982. *Ecology of Coastal Waters. A Systems Approach*. Blackwell, London.

Manson, F.J., Loneragan, N.R., Harch, B.D., Skilleter, G.A., and Williams, L. 2005. A broad-scale analysis of links between coastal fisheries production and mangrove extent: a case-study for northeastern Australia. *Fisheries Research* **74**: 69–85.

Marbà, N., Hemminga, M.A., Mateo, M.A., Duarte, C.M., Mass, Y.E.M., Terrados, J., and Gacia, E. 2002. Carbon and nitrogen translocation between seagrass ramets. *Marine Ecology Progress Series* **226**: 287–300.

Marguillier, S., van der Velde, G., Dehairs, F., Hemminga, M.A., and Rajagopal, S. 1997. Trophic relationships in an interlinked mangrove-seagrass ecosystem as traced by $\delta^{13}C$ and $\delta^{15}N$. *Marine Ecology Progress Series* **151**: 115–121.

Mazzotti, F.J. and Dunson, W.A. 1984. Adaptations of *Crocodylus acutus* and *Alligator* for life in saline water. *Comparative Biochemistry and Physiology* **79A**: 641–646.

McClanahan, T.R., Nugues, M., and Mwachireya, S. 1994. Fish and sea urchin herbivory and competition in Kenyan coral reef lagoons: the role of reef management. *Journal of Experimental Marine Biology and Ecology* **184**: 237–254.

McConchie, C.A. and Knox, R.B. 1989. Pollination and reproductive biology of seagrasses. In *Biology of Seagrasses. A Treatise on the Biology of Seagrasses with Special Reference to the Australian Region* (Larkum, A.W.D., McComb, A.J., and Shepherd, S.A., eds), pp. 74–111. Elsevier, Amsterdam.

McCoy, E.D., Mushinsky, H.R., Johnson, D., and Meshaka, W.E. 1996. Mangrove damage caused by Hurricane Andrew on the southwestern coast of Florida. *Bulletin of Marine Science* **59**: 1–8.

McGrady-Steed, J., Harris, P.M., and Morin, P.J. 1997. Biodiversity regulates ecosystem predictability. *Nature* **390**: 162–165.

McGuinness, K.A. 1997. Dispersal, establishment and survival of *Ceriops tagal* propagules in a north Australian mangrove forest. *Oecologia* **109**: 80–87.

McIvor, C.C. and Smith, T.J. 1996. Differences in the crab fauna of mangrove areas at southwest Florida and a northeast Australia location: implications for leaf litter processing. *Estuaries* **18**: 591–597.

McKee, K.L. 1993. Soil physicochemical patterns and mangrove species distribution—reciprocal effects? *Journal of Ecology* **81**: 477–487.

McKee, K.L. 1995. Mangrove species distribution and propagule predation in Belize: an exception to the dominance-predation hypothesis. *Biotropica* **27**: 334–345.

McKee, K.L. 2001. Root proliferation in decaying roots and old root channels: a nutrient conservation mechanism in oligotrophic mangrove forests? *Journal of Ecology* **89**: 876–887.

McKee, K.L. and Mendelssohn, I.A. 1987. Root metabolism in the Black Mangrove (*Avicennia germinans* (L.): response to hypoxia. *Environmental and Experimental Botany* **27**: 147–156.

McKenzie, N.L. and Rolfe, J.K. 1986. Structure of bat guilds in the Kimberley mangroves, Australia. *Journal of Animal Ecology* **55**: 401–420.

McMillan, C. 1975. Adaptive differentiation to chilling in mangrove populations. In *Proceedings of the International Symposium on Biology and Management of Mangroves* (Walsh, G.E., Snedaker, S.C., and Teas, H.J., eds), pp. 62–68. University of Florida, Gainesville, FL.

Micheli, F., Gherardi, F., and Vannini, M. 1991. Feeding and burrowing ecology of two East African mangrove crabs. *Marine Biology* **111**: 247–254.

Miller, D.C. 1961. The feeding mechanisms of fiddler crabs, with ecological considerations of feeding adaptations. *Zoologica* **46**: 89–100.

Milliman, J.D., Quraishee, G.S., and Beg, M.A.A. 1984. Sediment discharge from the Indus River to the Ocean: past, present and future. In *Marine Geology and Oceanography of Arabian Sea and Coastal Pakistan* (Haq, B.U. and Milliman, J.D., eds), pp. 65–70. Van Nostrand Reinhold, New York.

Moon, G.J., Clough, B.F., Peterson, C.A., and Allaway, W.G. 1986. Apoplastic and symplastic pathways in *Avicennia marina* (Forsk.) Vierh. roots revealed by fluorescent tracer dyes. *Australian Journal of Plant Physiology* **13**: 637–648.

Moran, M.A., Wicks, R.J., and Hodson, R.E. 1991. Export of dissolved organic matter from a mangrove swamp ecosystem—evidence from natural fluorescence, dissolved lignin phenols, and bacterial secondary production. *Marine Ecology Progress Series* **76**: 175–184.

Mukai, H. 1993. Biogeography of the tropical seagrasses in the Western Pacific. *Australian Journal of Marine and Freshwater Research* **44**: 1–17.

Mumby, P.J., Edwards, A.J., Arias-González, J.E., Lindeman, K.C., Blackwell, P.G., Gall, A. *et al.* 2004. Mangroves enhance the biomass of coral reef fish in the Caribbean. *Nature* **427**: 533–536.

Muramatsu, Y., Harada, A., Ohwaki, Y., Kasahara, Y., Takagi, S., and Fukuhara, T. 2002. Salt-tolerant ATPase activity in the plasma membrane of the marine angiosperm *Zostera marina* L. *Plant Cell Physiology* **43**: 1137–1145.

Naeem, S. and Li, S. 1997. Biodiversity enhances ecosystem reliability. *Nature* **390**: 507–509.

Nagelkerken, I. and van der Velde, G. 2004. Relative importance of interlinked mangroves and seagrass beds as feeding habitats for juvenile reef fish on a Caribbean island. *Marine Ecology Progress Series* **274**: 153–159.

Nakaoka, M. 2005. Plant-animal interactions in seagrass beds: ongoing and future challenges for understanding population and community dynamics. *Population Ecology* **47**: 167–177.

Neilson, M.J., Giddins, R.L., and Richards, G.N. 1986. Effects of tannins on the palatability of mangrove leaves to the tropical sesarmid crab *Neosarmatium smithi*. *Marine Ecology Progress Series* **34**: 185–186.

Nielsen, M.G. 1997. Nesting biology of the mangrove mud-nesting ant *Polyrhachis sokolova* Forel (Hymenoptera, Formicidae) in northern Australia. *Insectes Sociaux* **44**: 15–21.

Noske, R.A. 1995. The ecology of mangrove forest birds in Peninsular Malaysia. *Ibis* **137**: 250–263.

Noske, R.A. 1996. Abundance, zonation and foraging ecology of birds in mangroves of Darwin Harbour, Northern Territory. *Wildlife Research* **23**: 443–474.

Odum, W.E. and Heald, E.J. 1975. The detritus-based food web of an estuarine mangrove community. *Estuarine Research* **1**: 265–286.

Offenberg, J., Nielsen, M.G., Macintosh, D.J., Havanon, S., and Aksornkoae, S. 2004a. Lack of ant attendance may induce compensatory plant growth. *Oikos* **111**: 170–178.

Offenberg, J., Havanon, S., Aksornkoae, S., Macintosh, D.J., and Nielsen, M.G. 2004b. Observations on the ecology of weaver ants (*Oecophylla smaragdina* Fabricius) in a Thai mangrove ecosystem and their effect on herbivory of *Rhizophora mucronata* Lam. *Biotropica* **36**: 344–351.

Offenberg, J., Nielsen, M.G., Macintosh, D.J., Havanon, S., and Aksornkoae, S. 2004c. Evidence that insect herbivores are deterred by ant pheromones. *Proceedings of the Royal Society of London Series B* **271**: S433–S435.

Ólafsson, E. and Ndaro, S.G.M. 1997. Impact of the mangrove crabs *Uca annulipes* and *Dotilla fenestrata* on meiobenthos. *Marine Ecology Progress Series* **158**: 225–232.

Ólafsson, E., Buchmayer, S., and Skov, M.W. 2002. The East African Decapod crab *Neosarmatium meinerti* (de Man) sweeps mangrove floors clean of leaf litter. *Ambio* **31**: 569–573.

Ong, J.E. 1993. Mangroves—a carbon source and sink. *Chemosphere* **27**: 1097–1107.

Ong, J.E. 1995. The ecology of mangrove conservation and management. *Hydrobiologica* **295**: 343–351.

Ong, J.E., Gong, W.K., Wong, C.H., and Dhanarajan, G. 1984. Contribution of aquatic productivity in managed mangrove ecosystem in Malaysia. In *Proceedings of the Asian Symposium on Mangrove Environment Research and Management* (Soepadmo, E., Rao, A.N., and Macintosh, D.J., eds), pp. 209–215. University of Malaya, Kuala Lumpur.

Ono, Y. 1965. On the ecological distribution of ocypodid crabs in the estuary. *Memoirs of the Faculty of Science of Kyushu University* **4**: 1–60.

Onuf, C.P., Teal, J.M., and Valiela, I. 1977. Interactions of nutrients, plant growth and herbivory in a mangrove ecosystem. *Ecology* **58**: 514–526.

Osborne, D.J. and Berjak, P. 1997. The making of mangroves: the remarkable pioneering role played by the seed of *Avicennia marina*. *Endeavour* **21**: 143–147.

Osborne, K. and Smith, T.J. 1990. Differential predation on mangrove propagules in open and closed canopy forest habitats. *Vegetatio* **89**: 1–6.

Pang, S.C. 1989. *Traditional Fishing Activities in the Mangrove Ecosystems of Sarawak*. Department of Fisheries, Kuala Lumpur.

Pannier, F. and Fraíno de Pannier, R. 1975. Physiology of vivipary in *Rhizophora mangle*. In *Proceedings of the International Symposium on Biology and Management of Mangroves* (Walsh, G.E., Snedaker, S.C., and Teas, H.J., eds), pp. 632–639. University of Florida, Gainesville, FL.

Parani, M., Lakshmi, M., Elango, S., Ram, N., Anuratha, C.S., and Parida, A. 1997. Molecular phylogeny of mangroves II. Intra- and inter-specific variation in *Avicennia* revealed by RAPD and RFLO markers. *Genome* **40**: 487–495.

Passioura, J.B., Ball, M.C., and Knight, J.H. 1992. Mangroves may salinize the soil and in so doing limit their transpiration rate. *Functional Ecology* **6**: 476–481.

Pernetta, J.C. 1993. *Mangrove Forests, Climate Change and Sea Level Rise: Hydrological Influences on Community Structure and Survival, with Examples from the Indo-West Pacific*. International Union for the Conservation of Nature, Gland.

Perry, D.M. 1988. Effects of associated fauna on growth and productivity in the red mangrove. *Ecology* **6**: 1064–1075.

Peters, E.C., Gassman, N.J., Firman, J.C., Richmond, R.H., and Power, E.A. 1997. Ecotoxicology of tropical marine ecosystems. *Environmental Toxicology and Chemistry* **16**: 12–40.

Pidcock, S., Taplin, L.E., and Grigg, G.C. 1997. Differences in renal-cloacal function between *Crocodylus porosus* and *Alligator mississippiensis* have implications for crocodilian evolution. *Journal of Comparative Physiology B* **167**: 153–158.

Pittman, S.J., McAlpine, C.A., and Pittman, K.M. 2004. Linking fish and prawns to their environment: a hierarchical landscape approach. *Marine Ecology Progress Series* **283**: 233–254.

Plaziat, J.C. 1984. Mollusk distribution in the mangal. In *Hydrobiology of the Mangal* (Por, F.D. and Dor, I., eds), pp. 111–143. W. Junk, The Hague.

Plaziat, J.C. 1995. Modern and fossil mangroves and mangals: their climatic and biogeographic variability. In *Marine Palaeoenvironmental Analysis from Fossils* (Bosence, D.W.J. and Allison, P.A., eds), Geological Society Special Publication vol. 83, pp. 73–96. Geological Society, London.

Plaziat, J.-C., Cavagnetto, C., Koeniguer, J.-C., and Baltzer, F. 2001. History and biogeography of the mangrove ecosystem, based on a critical reassessment of the paleontological record. *Wetlands Ecology and Management* **9**: 161–179.

Preen, A. 1995. Impacts of dugong foraging on seagrass habitats: observational and experimental evidence for cultivation grazing. *Marine Ecology Progress Series* **124**: 201–213.

Price, A.R.G., Medley, P.A.H., McDowall, R.J., Dawson-Shepherd, A.R., Hogarth, P.J., and Ormond, R.F.G. 1987. Aspects of mangal ecology along the Red Sea coast of Saudi Arabia. *Journal of Natural History* **21**: 449–464.

Primavera, J.H. 1997a. Fish predation on mangrove-associated penaeids. The role of structures and substrate. *Journal of Experimental Marine Biology and Ecology* **215**: 205–216.

Primavera, J.H. 1997b. Socio-economic impacts of shrimp culture. *Aquaculture Research* **28**: 815–827.

Putz, F.E. and Chan, H.-T. 1986. Tree growth, dynamics and productivity in a mature mangrove forest in Malaysia. *Forest Ecology and Management* **17**: 211–230.

Qureshi, M.T. 1996. Restoration of mangroves in Pakistan. In *Restoration of Mangrove Ecosystems* (Field, C., ed), pp. 126–142. International Timber Organisation and International Society for Mangrove Ecosystems, Okinawa.

Rabinowitz, D. 1978a. Early growth of mangrove seedlings in Panama, and an hypothesis concerning the relationship of dispersal and zonation. *Journal of Biogeography* **5**: 113–133.

Rabinowitz, D. 1978b. Dispersal properties of mangrove propagule. *Biotropica* **10**: 47–57.

Rabinowitz, D. 1978c. Mortality and initial propagule size in mangrove seedlings in Panama. *Journal of Ecology* **66**: 45–61.

Randall, D.J. and Tsui, T.K.N. 2002. Ammonia toxicity in fish. *Marine Pollution Bulletin* **45**: 17–23.

Randall, J.E. 1965. Grazing effect on sea grasses by herbivorous reef fishes in the West Indies. *Ecology* **46**: 255–260.

Reusch, T.B.H. 2003. Floral neighbourhoods in the sea: how floral density, opportunity for outcrossing and population fragmentation affect seed set in *Zostera marina*. *Journal of Ecology* **91**: 610–615.

Reusch, T.B.H., Bostrom, C., and Stam, W.T. 1999. An ancient eelgrass clone in the Baltic. *Marine Ecology Progress Series* **183**: 301–304.

Reusch, T.B.H., Ehlers, A., Hämmerl, A., and Worm, B. 2005. Ecosystem recovery after climatic extremes enhanced by genotypic diversity. *Proceedings of the National Academy of Sciences USA* **102**: 2826–2831.

Ricklefs, R.E. and Latham, R.E. 1993. Global patterns of diversity in mangrove floras. In *Species Diversity in Ecological Communities* (Ricklefs, R.E. and Schluter, D., eds), pp. 215–229. University of Chicago Press, Chicago.

Ricklefs, R.E. and Schluter, D. 1993. Species diversity: regional and historical influences. In *Species Diversity in Ecological Communities* (Ricklefs, R.E. and Schluter, D., eds), pp 350–363. University of Chicago Press, Chicago.

Ridd, P.V. 1996. Flow through animal burrows in mangrove creeks. *Estuarine, Coastal and Shelf Science* **43**: 617–625.

Ridd, P.V. and Samm, R. 1996. Profiling groundwater salt concentrations in mangrove swamps and tropical salt flats. *Estuarine, Coastal and Shelf Science* **43**: 627–635.

Rivera Monroy, V.H., Day, J.W., Twilley, R.R., and Veraherrera, F. 1995. Flux of nitrogen and sediment in a fringe mangrove forest in Terminos Lagoon, Mexico. *Estuarine, Coastal and Shelf Science* **40**: 139–160.

Robertson, A.I. 1988. *Food Chains in Tropical Australian Mangrove Habitats: a Review of Recent Research*. UNDP/UNESCO Symposium on New Perspectives in Research and Management of Mangrove Ecosystems, New Delhi.

Robertson, A.I. 1991. Plant-animal interactions and the structure and function of mangrove forest ecosystems. *Australian Journal of Ecology* **16**: 433–443.

Robertson, A.I. and Duke, N.C. 1987. Insect herbivory on mangrove leaves in North Queensland. *Australian Journal of Ecology* **12**: 1–7.

Robertson, A.I. and Daniel, P.A. 1989. The influence of crabs on litter processing in high intertidal mangrove forests in tropical Australia. *Oecologia* **78**: 191–198.

Robertson, A.I. and Blaber, S.J.M. 1992. Plankton, epibenthos and fish communities. In *Tropical Mangrove Ecosystems* (Robertson, A.I. and Alongi, D.M., eds), pp. 173–224. American Geophysical Union, Washington DC.

Robertson, A.I. and Phillips, M.J. 1995. Mangroves as filters of shrimp pond effluent: prediction and biochemical research needs. *Hydrobiologia* **295**: 311–321.

Robertson, A.I., Giddins, R., and Smith, T.J. 1990. Seed predation by insects in tropical mangrove forests: extent and effects on seed viability and the growth of seedlings. *Oecologia* **83**: 213–219.

Robertson, A.I., Alongi, D.M., and Boto, K.G. 1992. Food chains and carbon fluxes. In *Tropical Mangrove Ecosystems* (Robertson, A.I. and Alongi, D.M., eds), pp. 293–326. American Geophysical Union, Washington DC.

Rodelli, M.R., Gearing, J.N., Gearing, P.J., Marshall, N., and Sasekumar, A. 1984. Stable isotope ratio as a tracer of mangrove carbon in Malaysian ecosystems. *Oecologia* **61**: 326–333.

Rollon, R.N., Vermaat, J.E., and Nacorda, H.M.E. 2003. Sexual reproduction in S.E. Asian seagrasses: the absence of a seed bank in *Thalassia hemprichii*. *Aquatic Botany* **75**: 181–185.

Rosenzweig, M.L. 1995. *Species Diversity in Space and Time*. Cambridge University Press, Cambridge.

Ruckelshaus, M.H. 1996. Estimation of genetic neighbourhood parameters from pollen and seed dispersal in the marine angiosperm *Zostera marina*, L. *Evolution* **50**: 856–864.

Saenger, P. 2002. *Mangrove Ecology, Silviculture and Conservation*. Kluwer, Dordrecht.

Saenger, P. and Snedaker, S.C. 1993. Pantropical trends in mangrove above-ground biomass and annual litterfall. *Oecologia* **96**: 293–299.

Saenger, P. and Bellan, M.F. 1995. *The Mangrove Vegetation of the Atlantic Coast of Africa. A Review*. Laboratoire d'Ecologie Terrestre, Toulouse.

Saintilan, N. 1997. Above- and below-ground biomasses of two species of mangrove on the Hawkesbury River estuary, New South Wales. *Marine and Freshwater Research* **48**: 147–152.

Salter, R.E., MacKenzie, N.A., Nightingale, N., Aken, K.M., and Chai, P.P.K. 1985. Habitat use, ranging behaviour, and food habits of the Proboscis monkey, *Nasalis larvatus* (van Wurmb), in Sarawak. *Primates* **26**: 436–451.

Sasekumar, A. 1994. Meiofauna of a mangrove shore on the west coast of Peninsular Malaysia. *Raffles Bulletin of Zoology* **42**: 901–915.

Sasekumar, A., Chong, C.V., Leh, M.U., and D'Cruz, R. 1992. Mangroves as a habitat for fish and prawns. *Hydrobiologia* **247**: 195–207.

Sasekumar, A., Ong, T.L., and Thong, K.L. 1984. Predation of mangrove fauna by marine fishes. In *Proceedings of the Asian Symposium on the Mangrove Environment—Research and Management* (Soepadmo, E., Rao, A.N., and Macintosh, D.J., eds), pp. 378–384. University of Malaya, Kuala Lumpur.

Scholander, P.F. 1955. How mangroves desalinate seawater. *Physiologia Plantarum* **21**: 251–261.

Scholander, P.F., van Dam, L., and Scholander, S.I. 1955. Gas exchange in the roots of mangroves. *American Journal of Botany* **42**: 92–98.

Schrijvers, J. and Vincx, M. 1997. Cage experiments in an East African mangrove forest: a synthesis. *Journal of Sea Research* **38**: 123–133.

Schrijvers, J., Schallier, R., Silence, J., Okondo, J.P., and Vincx, M. 1997. Interactions between epibenthos and meiobenthos in a high intertidal *Avicennia marina* mangrove forest. *Mangroves and Salt Marshes* **1**: 137–154.

Short, F.T. and Neckles, H.A. 1999. The effects of global climate change on seagrasses. *Aquatic Botany* **63**: 169–196.

Short. F.T. and Willey-Echevarria, S. 1996. Natural and human-induced disturbance of seagrasses. *Environmental Conservation* **23**: 17–27.

Silberstein, K., Chiffings, A.W., and McComb, A.J. 1986. The loss of seagrass in Cockburn Sound, Western Australia. III. The effect of epiphytes on production of *Posidonia australis* Hook. F. *Aquatic Botany* **24**: 355–371.

Simberloff, D. 1976. Experimental zoogeography of islands: effects of island size. *Ecology* **57**: 629–648.

Simberloff, D.S. 1983. Mangroves. In *Costa Rican Natural History* (Janzen, D.H., eds), pp. 273–276. University of Chicago Press, Chicago.

Simberloff, D.S. and Wilson, E.O. 1969. Experimental zoogeography of islands: the colonization of empty islands. *Ecology* **50**: 278–296.

Simberloff, D.S. and Wilson, E.O. 1970. Experimental zoogeography of islands: a two-year record of colonization. *Ecology* **51**: 934–937.

Simberloff, D., Brown, B.J., and Lowrie, S. 1978. Isopod and insect root borers may benefit Florida mangroves. *Science* **201**: 630–632.

Skelton, N.J. and Allaway, W.G. 1996. Oxygen and pressure changes measured in situ during flooding in roots of the Grey Mangrove *Avicennia marina* (Forssk.) Vierh. *Aquatic Botany* **54**: 165–175.

Skov, M.W. and Hartnoll, R.G. 2002. Paradoxical selective feeding on a low-nutrient diet: why do mangrove crabs eat leaves? *Oecologia* **131**: 1–7.

Slim, F.J., Hemminga, M.A., Ochieng, C., Jannink, N.T., Cocheret de la Morinière, E., and van der Velde, G. 1997. Leaf litter removal by the snail *Terebralia palustris* (Linnaeus) and sesarmid crabs in an East African mangrove forest. *Journal of Experimental Marine Biology and Ecology* **215**: 35–48.

Smith, T.J. 1987a. Effects of light and intertidal position on seeding survival and growth in tropical tidal forests. *Journal of Experimental Marine Biology and Ecology* **110**: 133–146.

Smith, T.J. 1987b. Effects of seed predators and light level on the distribution of *Avicennia marina* (Forsk.) Vierh. in tropical, tidal forests. *Estuarine, Coastal and Shelf Science* **25**: 43–51.

Smith, T.J. 1987c. Seed predation in relation to tree dominance and distribution in mangrove forests. *Ecology* **68**: 266–273.

Smith, T.J. 1992. Forest structure. In *Tropical Mangrove Ecosystems* (Robertson, A.I. and Alongi, D.M., eds), pp. 101–136. American Geophysical Union, Washington DC.

Smith, T.J., Boto, K.G., Frusher, S.D., and Giddins, R.L. 1991. Keystone species and mangrove forest dynamics: the influence of burrowing by crabs on soil nutrient status and forest productivity. *Estuarine, Coastal and Shelf Science* **33**: 19–32.

Snedaker, S.C. 1982. Mangrove species zonation: why? In *Contributions to the Study of Halophytes*. Tasks for Vegetation Science vol. 2 (Sen, D.N. and Rajpurohit, K.S., eds), pp. 111–125. W. Junk, The Hague.

Snedaker, S.C., Baquer, S.J., Behr, P.J., and Ahmed, S.I. 1995. Biomass distribution in *Avicennia marina* plants in the Indus River Delta, Pakistan. In *The Arabian Sea: Living Marine Resources and the Environment* (Thompson, M.F. and Tirmizi, N.M., eds), pp. 389–394. Balkema, Rotterdam.

Sodhi, N.S., Choo, J.P.S., Lee, B.P.Y.-H., Quek, K.C., and Kara, A.U. 1997. Ecology of a mangrove forest bird community in Singapore. *Raffles Bulletin of Zoology* **45**: 1–13.

Sogard, S.M. 1992. Variability in growth rates of juvenile fishes in different estuarine habitats. *Marine Ecology Progress Series* **85**: 35–53.

Somerfield, P.J., Gee, J.M., and Aryuthaka, C. 1998. Meiofaunal communities in a Malaysian mangrove forest. *Journal of the Marine Biological Association of the United Kingdom* **78**: 712–732.

Sousa, W.P., Kennedy, P.G., and Mitchell, B.J. 2003. Propagule size and predispersal damage by insects affect establishment and early growth of mangrove seedlings. *Oecologia* **135**: 564–575.

Spalding, M., Blasco, F., and Field, C. (eds) 1997. *World Mangrove Atlas*. International Society for Mangrove Ecosystems, Okinawa.

Spaninks, F. and van Beukering, P. 1997. *Economic Valuation of Mangrove Ecosystems: Potential and Limitations*. CREED Working Paper 14, pp. 1–53. International Institute for Environment and Development, Amsterdam.

Stafford-Deitsch, J. 1996. *Mangrove. The Forgotten Habitat*. Immel, London.

Stapel, J., Aarts, T.L., van Duynhoven, B.H.M., de Groot, J.D., van den Hoogen, P.H.W., and Hemminga, M.A. 1996. Nutrient uptake by leaves and roots of the seagrass *Thalassia hemprichii* in the Spermonde Archipelago, Indonesia. *Marine Ecology Progress Series* **134**: 195–206.

Start, A.N. and Marshall, A.G. 1976. Nectarivorous bats as pollinators of trees in West Malaysia. In *Tropical Trees. Variation, Breeding and Conservation* (Burley, J. and Styles, B.T., eds), pp. 141–150. Academic Press, London.

Steinke, T.D. 1975. Some factors affecting dispersal and establishment of propagules of *Avicennia marina* (Forsk.) Vierh. In *Proceedings of the International Symposium on Biology and Management of Mangroves* (Walsh, G.E., Snedaker, S.C., and Teas, H.J., eds), pp. 402–414. University of Florida, Gainesville, FL.

Steinke, T.D. and Ward, C.J. 1988. Litter production by mangroves. II. St Lucia and Richards Bay. *South African Journal of Botany* **54**: 445–454.

Steinke, T.D., Barnabas, A.D., and Somaru, R. 1990. Structural changes and associated microbial activity accompanying decomposition of mangrove leaves in Mgeni Estuary. *South African Journal of Botany* **56**: 39–49.

Sternberg, L.da S.L. and Swart, P.K. 1987. Utilization of freshwater and ocean water by coastal plants of southern Florida. *Ecology* **68**: 1898–1905.

Stoner, A.W., Ray, M., and Waite, J.M. 1995. Effects of a large herbivorous gastropod on macrofauna communities in tropical seagrass meadows. *Marine Ecology Progress Series* **121**: 125–137.

Sussex, K. 1975. Growth and metabolism of the embryo and attached seedling of the viviparous mangrove, *Rhizophora mangle*. *American Journal of Botany* **62**: 948–953.

Sutherland, J.P. 1980. Dynamics of the epibenthic community on roots of the mangrove *Rhizophora mangle*, at Bahia de Buche, Venezuela. *Marine Biology* **58**: 75–84.

Svavarsson, J., Osore, M.K.W., and Olafsson, E. 2002. Does the wood-borer *Sphaeroma terebrans* (Crustacea) shape the distribution of the mangrove *Rhizophora mucronata*? *Ambio* **31**: 574–579.

Takeda, S., Matsumasa, M., Kikuchi, S., Poovachiranon, S., and Murai, M. 1996. Variation in the branchial formula of semiterrestrial crabs (Decapoda: Brachyura: Grapsidae and Ocypodidae) in relation to physical adaptations to the environment. *Journal of Crustacean Biology* **16**: 472–486.

Taplin, K.A., Irlandi, E.A., and Raves, R. 2005. Interference between the macroalga *Caulerpa prolifera* and the seagrass *Halodule wrightii*. *Aquatic Botany* **83**: 175–186.

Terrados, J. and Williams, S.L. 1997. Leaf versus root nitrogen uptake by the surfgrass *Phyllospadix torreyi*. *Marine Ecology Progress Series* **149**: 267–277.

Terrados, J., Duarte, C.M., Kamp-Nielsen, L., Agawin, N.S.R., Gacia, E., Lacap, D. et al. 1999. Are seagrass growth and survival constrained by the reducing conditions of the sediment? *Aquatic Botany* **65**: 175–197.

Thayer, G.W., Engel, D.W., and Bjorndal, K.A. 1982. Evidence for short-circuiting of the detritus cycle of seagrass beds by the Green turtle, *Chelonia mydas* L. *Journal of Experimental Marine Biology and Ecology* **62**: 173–183.

Thayer, G.W., Bjorndal, K.A., Ogden, J.C., Williams, S.L., and Zieman, J.C. 1984. Role of larger herbivores in seagrass communities. *Estuaries* **7.4A**, 351–376.

Thibodeau, F.R. and Nickerson, N.H. 1986. Differential oxidation of mangrove substrate by *Avicennia germinans* and *Rhizophora mangle*. *American Journal of Botany* **73**: 512–516.

Thom, B.G. 1984. Coastal landforms and geomorphic processes. In *The Mangrove Ecosystem: Research Methods* (Snedaker, S.C. and Snedaker, J.G., eds), pp. 3–17. UNESCO, Paris.

Thongtham, N. and Kristensen, E. 2005. Carbon and nitrogen balance of leaf-eating sesarmid crabs (*Neoepisesarma versicolor*) offered different food sources. *Estuarine, Coastal and Shelf Science* **65**: 213–222.

Tilman, D. 1996. Biodiversity: population versus ecosystem stability. *Ecology* **77**: 350–363.

Tomlinson, P.B. 1986. *The Botany of Mangroves*. Cambridge University Press, Cambridge.

Tuan, M.S., Ninomiya, I., and Ogino, K. 1995. Salt uptake and excretion in the mangrove. *Avicennia marina* (Forsk.) Vierh. *Tropics* **5**: 69–79.

Twilley, R.R. 1985. The exchange of organic carbon in basin mangrove forests in a southwest Florida estuary. *Estuarine, Coastal and Shelf Science* **20**: 543–557.

Tyerman, S.D. 1989. Solute and water relations of seagrasses. In *Biology of Seagrasses. A Treatise on the Biology of Seagrasses with Special Reference to the Australian Region* (Larkum, A.W.D., McComb, A.J., and Shepherd, S.A., eds), pp. 723–759. Elsevier, Amsterdam.

Udy, J.W. and Dennison, W.C. 1997. Growth and physiological responses of three seagrass species to elevated sediment nutrients in Moreton Bay, Australia. *Journal of Experimental Marine Biology and Ecology* **217**: 253–277.

Valiela, I., Bowen, J.L., and York, J.K. 2001. Mangrove forests: one of the world's threatened major tropical environments. *BioScience* **51**: 807–815.

van der Valk, A.G. and Attiwill, P.M. 1984. Decomposition of leaf and root litter of *Avicennia marina* at Westernport Bay, Victoria, Australia. *Aquatic Botany* **18**: 205–221.

Vanhove, S., Vincx, M., van Gansbeke, D., Gijselinck, W., and Schram, D. 1992. The meiobenthos of five mangrove vegetation types in Gazi Bay, Kenya. *Hydrobiologia* **247**: 99–108.

Vannini, M. and Ruwa, R.K. 1994. Vertical migrations in the tree crab *Sesarma leptosoma* (Decapoda, Grapsidae). *Marine Biology* **118**: 271–278.

Veenakumari, K., Mohanraj, P., and Bandyopadhyay, A.K. 1997. Insect herbivores and their natural enemies in the mangals of the Andaman and Nicobar islands. *Journal of Natural History* **31**: 1105–1126.

Vermaat, J.E., Rollon. R.N., Lacap, C.D.A., Billot, C., Alberto, F., Nacorda, H.M.E. et al. 2004. Meadow fragmentation and reproductive output of the S.E. Asian seagrass *Enhalus acoroides*. *Journal of Sea Research* **52**: 321–328.

Vermeij, G.J. 1974. Molluscs in mangrove swamps: physiognomy, diversity, and regional differences. *Systematic Zoology* **22**: 609–624.

Voris, H.K. and Jeffries, W.B. 1995. Predation on marine snakes—a case for decapods supported by new observations from Thailand. *Journal of Tropical Ecology* **11**: 569–576.

Walker, D.I. 1989. Regional studies—seagrass in Shark Bay, the foundations of an ecosystem. In *Biology of Seagrasses. A Treatise on the Biology of Seagrasses with Special Reference to the Australian Region* (Larkum, A.W.D., McComb, A.J., and Shepherd, S.A., eds), pp. 182–210. Elsevier, Amsterdam.

Wang, W.Q., Wang, M., and Lin, P. 2003. Seasonal changes in element contents in mangrove retranslocation during leaf senescence. *Plant and Soil* **252**: 187–193.

Warner, G. 1977. *Biology of Crabs*. Elek, London.

Warren, J.H. and Underwood, A.J. 1986. Effects of burrowing crabs on the topography of mangrove swamps in New South Wales. *Journal of Experimental Marine Biology and Ecology* **102**: 223–235.

Watson, J.G. 1928. Mangrove forests of the Malay Peninsula. *Malayan Forest Records* **6**: 1–275.

Watson, R.T., Albritton, D.L., Barker, T., Bashmakov, I.A., Canziani, O., Christ, R. et al. 2001. *Climate Change 2001: Synthesis Report*. Intergovernmental Panel on Climate Change, Geneva.

Weissburg, M. 1993. Sex and the single forager—gender-specific energy maximization strategies in fiddler crabs. *Ecology* **74**: 279–291.

Whitten, A.J. and Damanik, S.J. 1986. Mass defoliation of mangroves in Sumatra, Indonesia. *Biotropica* **18**: 176.

Wilson, K.A. 1989. Ecology of mangrove crabs—predation, physical factors and refuges. *Bulletin of Marine Science* **44**: 263–273.

Wolanski, E. 1995. Transport of sediment in mangrove swamps. *Hydrobiologia* **295**: 31–42.

Wolanski, E. and Chappell, J. 1996. The response of tropical Australian estuaries to a sea level rise. *Journal of Marine Systems* **7**: 267–279.

Woodroffe, C. 1987. Pacific island mangroves: distribution and environmental settings. *Pacific Science* **41**: 166–185.

Woodroffe, C.D. 1988. Relict mangrove stand on a Last Interglacial terrace, Christmas Island, Indian Ocean. *Journal of Tropical Ecology* **4**: 1–17.

Woodroffe, C. 1992. Mangrove sediments and geomorphology. In *Tropical Mangrove Ecosystems* (Robertson, A.I. and Alongi, D.M., eds), pp. 7–41. American Geophysical Union, Washington DC.

Woods, C.M.C. and Schiel, D.R. 1997. Use of seagrass *Zostera novazelandica* (Setchell, 1993) as habitat and food by the crab *Macrophthalmus hirtipes* (Heller, 1862) (Brachyura: Ocypodidae) on rocky intertidal platforms in southern New Zealand. *Journal of Experimental Marine Biology and Ecology* **214**: 49–65.

Wyn Jones, R.B. and Storey, R. 1981. Betaines. In *Physiology and Biochemistry of Drought Resistance in Plants* (Paleg, L.G. and Aspinall, D., eds), pp. 171–204. Academic Press, Sydney.

Yeager, C.P. 1989. Feeding ecology of the Proboscis monkey (*Nasalis larvatus*). *International Journal of Primatology* **10**: 497–530.

Young, B.M. and Harvey, L.E. 1996. A spatial analysis of the relationship between mangrove (*Avicennia marina* var. *australoasica*) physiognomy and sediment accretion in the Hauraki Plains, New Zealand. *Estuarine, Coastal and Shelf Science* **42**: 231–246.

Zimmerman, R.C., Kohrs, D.G., and Alberte, R.S. 1996. Top-down impact through a bottom-up mechanism: the effect of limpet grazing on growth, productivity and carbon allocation of *Zostera marina* L. (eelgrass). *Oecologia* ***107***: 560–587.

Zuberer, D.A. and Silver, W.S. 1975. Mangrove associated nitrogen fixation. In *Proceedings of the International Symposium on Biology and Management of Mangroves* (Walsh, G.E., Snedaker, S.C., and Teas, H.J., eds), pp. 643–653. University of Florida, Gainesville, FL.

Index

abscisic acid (ABA) 30–1
Acanthus 20, 29, 33, 61
accretion 60, 64, 92
   see also sedimentation
acetylene 25
Acrostichum 2, 56, 61–2, 126, 203, 212
   aureum 187
   speciosum 74
Adinia 160
Aëdes:
   alternans 80
   amesii 80
   vigilax 80
Aegialitis 2, 16, 20
Aegiceras 2, 16, 18–19, 20–2, 25, 111, 225
   community structure and dynamics 55, 66
   corniculatum 74, 109
   reproductive adaptations 29, 31
   terrestrial components of mangroves 73, 82
aerenchyma 14–15
aerial roots 9, 11, 14, 17, 23–4, 99
aerobic conditions 15
Africa 92, 185, 189, 196
   east 109, 113, 128, 136, 186–7
   north 189
   west 87, 113, 128, 186, 187–9, 194, 209
agriculture 235
Alaska 145
alcohol 209
   dehydrogenase 16
algae 65, 98–9, 106, 137, 138, 222
   see also microalgae
alligators 87–8, 161
alpheid shrimps 140
Alpheus (pistol shrimp) 125
   edamensis 140
Amazon delta 50, 161
Americas 189
ammonia 25, 45, 136
ammonium 27, 29, 67, 126, 171
amphibians 83–5
Amphibolis 5, 7, 46–8, 70
   antarctica 45, 164

amphipods 102, 132, 138, 140, 160
Amyema thalassium (mistletoe) 72
Anadara granosa (blood cockle) 213
anaerobic conditions 8, 12, 15–17, 58, 62, 120, 129, 168
Anas:
   acuta (pintail) 145
   penelope (wigeon) 145
anchorage 16–17
anchovies 133
Andaman island 75
animals from the land 73–97
   spiders 82–3
   see also insects; vertebrates
annelids *see* arachiannelids; oligochaetes; polychaetes; sabellid; serpulid
Annobo 194–5
anoxia *see* anaerobic conditions; redox potential
ant-house plants 72–3, 79
antelope 94
ants 77–9
Apis mellifera (honey bee) 82
Apocrytes (mudskipper) 135
aquaculture 178, 213–15, 234, 235
Arabia 94
   see also Sinai
Aratus pisonii 104–5, 113
Archaeozostera 200
archerfish 133
archiannelids 132
Argyrodes (spiders) 83
Armases elegans 113
arthropods 199
   see also crustacea; insects; spiders
ascidians 100
Asia 49, 93, 189, 209, 214
   central 91
   see also south-east Asia
Atlantic Ocean 4, 36, 100, 102, 125, 143, 168, 182, 186, 231
   north 6, 7
   south 7
   west 101, 141, 186–8

Atlantic-Caribbean-East Pacific region 185–7, 192–4, 203
ATP 16, 43
Atta (leafcutter ant) 77, 79
Australasia 186–8
Australia 24, 26
　biodiversity and biogeography 187, 189, 202
　Carpentaria, Gulf of 228
　community structure and dynamics of mangroves 51, 53
　Corner Inlet (Victoria) 4
　Darwin 90
　Hawkesbury River (New South Wales) 22
　impacts 209, 221, 231
　Kimberley (Western Australia) 95–6
　marine components of mangroves 112–13, 117, 126
　measuring and modelling of interactions 152, 158
　Missionary Bay (Queensland) 73–4
　north-eastern 52, 182
　north-western 93
　northern 64–6, 150, 233
　Queensland 93, 107, 110–11, 133, 155
　seagrass communities 144
　Shark Bay 70
　south 7
　south-eastern 34
　south-western 179
　terrestrial components of mangroves 72, 76, 79–80, 83, 85–7, 91–2, 94, 109
　tropical 77, 154
　western 68
　see also Hinchinbrook Island, Queensland
Avicennia 2, 10, 14–22, 24, 29, 31, 33, 35–7
　alba 197
　bicolor 36, 54
　biodiversity and biogeography 187, 189, 190, 197
　community structure and dynamics 51–2, 55–6, 60, 65–6
　germinans 25, 35–7, 54, 113, 196, 220
　impacts 220, 221
　marina 4, 12, 14, 18, 33–8
　　biodiversity and biogeography 187, 197
　　community structure and dynamics 55, 57, 61
　　comparisons and connections between mangroves and seagrasses 167
　　impacts 225, 231
　　marine components of mangroves 109, 111, 113
　　measuring and modelling of interactions 150
　　terrestrial components of mangroves 74, 76
　marine components of mangroves 98–9, 103, 110, 132
　measuring and modelling of interactions 151, 153, 156, 159
　officinalis 197
　terrestrial components of mangroves 72–6, 82, 91, 94–5
Avicenniaceae 2, 3
　see also Avicennia
Axis axis (axis deer) 94
Aythya americana (redhead duck) 145

bacteria 25, 65, 99, 137, 144, 156, 157, 159, 174
　see also cyanobacteria
Bahamas 139, 172, 196
Bahía las Minas 218, 220
Balanus spp. (barnacles) 99
Baltic Sea 41, 205
Bangladesh 161
　see also Sundarbans
bark 209
barnacles 65, 100, 175, 222
　see Balanus
basin mangroves 50–1
bats 95–6
bees 29, 82, 210
beetles *see* chrysomelid; *Coccotrypes*; fireflies; lampyrid; scolytid
Belize 73, 75, 101, 113, 178
biodiversity and biogeography 100, 183–205, 235
　definition of biodiversity 183–4
　diversity and ecosystem function 202–5
　diversity of mangrove fauna 197–9
　genetic diversity of mangroves 196–7
　local diversity of mangroves 192–6
　origins of mangroves 188–92
　regional diversity in mangroves 184–8
　seagrasses 200–2
biogeography *see* biodiversity and biogeography
Bioko (Fernando Po) 194–6
biomass 148–51, 152, 157, 163, 164–5, 204
　ratio, below-ground/above-ground 22
　standing-crop 203
biotic factors 228
birds 29, 89–93, 118, 145–6
bivalve molluscs 100, 106–7, 130, 132, 139, 175, 209–10
　see also mussels; oysters; piddocks; shipworms; teredinid
Blue Revolution 214, 216
Boiga dendrophila (cat snake) 86
Boleophthalmus (mudskipper) 135–6
Bombacaceae 2
boring organisms *see* wood-borers
Borneo 95, 210
Bostrychia 99

bostrychietum 99
Brachydontes 160
Brachyura 102
    see also crabs
Brahmaputra delta 52, 59
Branta bernicla (Brent goose) 145, 205
Brazil 125, 189
bromeliads 71–2
Bruguiera 2, 10, 14, 20, 29, 52
    biodiversity and biogeography 188, 189
    gymnorrhiza 56–7, 61, 74, 109, 113
    marine components of mangroves 132
    measuring and modelling of interactions 158
    parviflora 19, 56, 61
    terrestrial components of mangroves 72, 92, 94
bryozoans 100
Bubalus (water buffalo) 94
Bubulcus ibis (cattle egrets) 89
buffalo 94
bugs (Hemiptera) see coccids; see also homopteran
burrows/burrowing 26, 110, 126, 134–5, 140, 158–9
butterflies 29, 82

Caesalpina (vine) 71
Caiman crocodylus (common caiman) 87, 161
Calamus erinaceus (palm) 71
Callinectes sapidus 105
Caloglossa 99
camels 94, 225
Cameroon 195
Camptostemon 2
carbohydrates 31, 156
carbon 21, 22, 106–7, 168
    assimilation see photosynthesis
    comparisons and connections between mangroves and seagrasses 169, 175
    dissolved organic 162, 170–2
    flow 158
    flux 177
    measuring and modelling of interactions 151, 156, 159, 161
    monoxide 18
    /nitrogen ratio 74–5, 109–10, 114, 118, 157
    particulate organic 162, 170–2
    see also carbon dioxide; stable isotopes
carbon dioxide 9, 13–15, 22, 23, 176, 229, 231
    atmospheric rise 229–30
    seagrass communities 41–2, 138
carbonate 51
Cardisoma (land crab) 104
    carnifex 109–10
Caribbean 7, 25, 28, 178

biodiversity and biogeography 186, 188, 200
    impacts 218, 220, 233
    marine components of mangroves 100, 102, 112–13, 128
    seagrass communities 142, 143, 144
    terrestrial components of mangroves 76, 83
    see also Atlantic-Caribbean-East Pacific region
Catanella 99
catch per unit effort (CPUE) 226
catfish 133
cattle 225
cellulase 104
cellulose 143, 156–7
    -digesting bacteria 99, 144
Central America 35–6, 87, 91–2, 94, 104, 128
Ceratopogonidae (midges) 80
Cerberus (colubrid snakes) 88
Cercopithecus sp. (vervet monkey) 95
Ceriops 2, 18, 20, 29, 33, 36, 225
    australis 30, 57
    biodiversity and biogeography 188, 189
    community structure and dynamics 51, 52, 56
    decandra 19
    marine components of mangroves 109, 132
    tagal 19, 74, 107–9, 113
    terrestrial components of mangroves 79, 91
Cerithidea (snail) 127, 129, 133
Cervus duvauceli (swamp deer) 94
Chaerephon jobensis (bat) 96
Chalinolobus gouldii (bat) 96
charcoal 211–13, 233
    see also Matang
Chelonia mydas (green turtle) 143
China 73, 150, 210, 214, 216
chloride ions 84, 86
chlorophyll 12
Chlorophyta 99
Christmas Island 52, 59
chrysomelid beetles 75, 78
ciliate protozoa 131–2
clams 139
Cleistocoeloma merguiensis 105
Clibanarius (hermit crab) 125
    panamensis 99
climate 207
    change 229–33
    see also global warming
climbers 71
coastal protection and mangroves 210
coccids 74
Coccotrypes (scolytid beetle) 76
cockroaches 82
column roots 12

Combretaceae 2
  see also Laguncularia; Lumnitzera
community structure and dynamics 49–70, 101
  seagrass meadows 67–70
  see also form of the forest: mangroves
comparisons and connections 166–82
  commuting and other movement 177–8
  distinctiveness of mangroves and seagrasses 166–7
  exports from mangroves 172–4
  interactions 169
  larval dispersal and return 174–5
  mangroves and salt marshes 168–9
  mangroves, seagrasses and coral reefs 175–7
  mangroves, seagrasses and fisheries 178–82
  outwelling 170–1
competition 57, 69
Congeria 160
connections see comparisons and connections
Conocarpus 77
copepods 140
coral reefs 51, 175–7, 228
cormorants 89
Costa Rica 33, 99, 118
crab-eating hawk 92
crabs 27, 102–24, 170–1
  biodiversity and biogeography 204
  burrows 66
  faeces 158
  herbivory 112–13
  impacts 212, 222, 234
  leaf eating 103–7
  marine components of mangroves 100, 111–12
  measuring and modelling of interactions 157–8, 160
  mud, physiology of living in 119–23
  reproductive adaptations 118–19
  seagrass communities 138, 140
  seedlings 110–12
  selective feeders 107–10
  stress 123–4
  tree-climbing 113–14
  see also in particular Uca; sesarmid
Crassostrea (oyster) 100
creepers 2
Cretaceous era 189, 200
crocodiles 87–8, 161
Crocodylus:
  acutus (American crocodile) 87, 88
  niloticus (nile crocodile) 87
  porosus (estuarine crocodile) 87, 161
crustacea 102–27, 132
  as ecosystem engineers 125–7
  seagrass communities 140
  see also amphipods; copepods; crabs; cumaceans; isopods; shrimps

cryptovivipary 31, 36
Cuba 94
cumaceans 132
currents 4, 228
cyanobacteria 45, 98, 148
Cygnus:
  atratus (black swan) 145
  cygnus (whooper swan) 145
Cymodocea 5, 7, 47, 200
  serrulata 45
Cymodoceaceae 5
Cymodoceoideae 5
Cyprinodon 160

Dasyprocta (agouti) 94
decollation 129
decomposition 170, 232
deer 94
defoliation 75–6, 215
deltas 49–50
dentrification 28
Derris (vine) 61, 71
desiccation 67
Desulfovibrio (bacterium) 25
detritus/detritivores 73, 139, 140, 141, 172, 181
  see also leaf litter
diameter of trunk at breast height (DBH) 148–50
diatoms 98–9, 139
Dicaeum hirundinaceum (mistletoe bird) 72
Dicyathifer (shipworm) 130
dinitrophenol 18
direct uses of mangroves 208–9
dispersal 34–6, 47–8, 200, 201
dissolved organic matter 177
distinctiveness of mangroves and seagrass communities 166–7
diversity 184, 204
DNA synthesis 30
dolphins 93
dredging 229
Drosophilididae 29
ducks 145
Dugong dugon 144
dugongs 69, 138, 143, 144–5, 165
Durio zibethinus (durian) 97
dwarf mangroves 24, 53, 151, 153, 225
dynamics see community structure and dynamics

echinoderms 140–1
  see also holothurian, sea urchins
ecological effects 226
economic value of mangroves and seagrasses 233–6
  see also aquaculture; eco-tourism; exploitation; fisheries; forestry/timber
ecosystem function 202–5

ecotourism 211, 227
effluent 215–16
egrets 89
Egretta thula (snowy egrets) 89
Egypt *see* Sinai
Ellobiidae 190
ellobiids 190
encrusting organisms 99
energy flow 158, 163
Enhalus 5, 7, 47, 200
    acoroides 41, 46–7
Enteromorpha 99
Eocene era 190, 192, 200
Eonycteris (fruit bat) 97
    spelaea 96
epibionts 99–101
epiphytes 2, 25, 71, 78–9, 137–8, 139, 140
Episesarma (crab) 126
    versicolor 78, 105, 113, 158
epizoites 25
erosion 59–60, 64, 92, 169, 210, 215, 226, 234–5
estimation of biomass 148
estuarine position 193–4
estuary 175
ethanol 16
    see also alcohol
ethylene 25
Eudocimus ruber (scarlet ibis) 89
eukaryotes 98
Euphorbiaceae 2
    see also Excoecaria
Europe 189, 190, 200
euryhaline species 42, 44, 57, 84
eutrophication 176, 229
evaporation 66
exclusion 19, 42
Excoecaria 2, 16, 19–20, 29, 72–3, 75
    agallocha 74
exploitation 206–7
    see also aquaculture; fisheries; forestry
export of organic matter *see* outwelling
exports from mangroves 172–4
extracellular enzymes 156

faeces 158
fauna 99–101, 197–9
fecundity 33–4
Felis viverrima (fish cat) 93
ferns 71
    see also Acrostichum
fibre 111
fiddler crabs *see Uca*
Fiji 235
filter feeding *see* barnacles; bivalve molluscs
fireflies, synchronously flashing 80–2, 213
firewood 225–6
fish 118, 133–6, 138, 141–2, 160, 182
fisheries 178–82, 210, 213, 226, 234, 235
    see also shrimps

flatworms *see* turbellarian flatworms
flavoglycans 108
flavolans 107–9
flocculation 64
Florida 25, 27–8, 171–2, 179, 196, 207
    Biscayne Bay 44
    community structure and dynamics 51, 60, 69
    impacts 211, 220, 221
    Keys 101, 197–9
    measuring and modelling of interactions 160, 161, 163
    red mangrove 11
    south-eastern 53
    south-western 112–13
    southern 17, 176
    terrestrial components of mangroves 77, 86
flowers/flowering 46
    see also pollination
flycatcher 90
fodder 225–6
food-webs 163–4
foraminifera 132
forestry/timber 208–9, 211–13, 235
form of the forest: mangroves 49–67
    difference from other forests 62–4
    mud 64–7
    see also species zonation
formalin 26
fossil evidence 189–91, 200, 233
fouling organisms 99–101, 222
fringe mangroves 208
fruits 209
functional groups 202
fungi 25, 65, 79, 137, 156–7, 159
Fusarium 157
future of mangroves and seagrasses 236–7

Ganges delta 52, 59
gaps 61, 111, 221
gas:
    exchange 42, 152
    leakage 26
    spaces 12, 14–15
    transport 39
Gasteracanthidae (spiders) 83
gastropods 106–7, 113, 127, 139, 190
    see also Cerithidea; Littoraria; Littorinidae; *Morula; Telescopium; Terebralia; Thais;* thaidid; ellobiid
gastrotrichs 131, 132
Gecarcinidae (land crabs) 104
Gecarcoidea (land crabs) 104
geese 145
gei wai system 214, 234
genetic diversity 196–7, 205
geographical distribution 178–9, 231–2
geometrid moth 78
geomorphological change 59
Germany 190

Gerygone 90–1
  chloronata 91
  laevigaster 91
  magnirostris 91
gills 136
girth at breast height (GBH) 148–50
global warming 230–2
glycinebetaine 19, 44
glycolysis 16
goats 225
Gobiosoma bosci (naked goby) 141
Goniopsis 102, 113
goods and services of mangroves *see* economic value
Gracilaria 98
grapsid crabs 106, 113, 167
Grapsidae 102, 120
  see also Goniopsis; *Metopograpsus*; sesarmid
Grapsoid crabs 102, 124
Grapsoidea 102
  see also grapsid, sesarmid, ocypodid
growth 24
  and nutrients, inorganic 27–9
  of seagrasses 39–41
guano 27, 69
guilds 90–1

habitat restoration 207
Haemulon sciurus (bluestriped grunt) 178
Haliclona (sponge) 25
Halodule 5, 7, 47, 200
  pinifolia 201
  uninervis 45, 70, 201
  wrightii 43–4, 48, 69–70, 145, 201
Halophila 4–5, 7, 41, 46, 48, 67, 144–5, 200
  decipiens 40
  johnsoni 48
  ovalis 67, 144
Halophiloideae 5
hammock mangroves 51
harpacticoid copepod 131, 132, 133
haustorium 72
heavy-metal contamination 217
Hemiramphus spp. (halfbeaks) 142
herbicides 217
herbivory 135, 138–42, 144, 164, 168, 170
  crabs 103, 112–13
  insects 73–7
Heritiera 2, 59, 73, 76
  littoralis 74
hermit crabs 102, 125
herons 89, 118
Herpestes spp. (mongooses) 93
Heterozostera 5, 7
Hinchinbrook Island, Queensland 27–9, 132, 170–1, 174
Holocene era 51, 233
holothurian sea cucumbers 141
Holothuria spp. (bêche-de-mer/trepang) 141

homopteran bugs 75
Honduras 211
honeyeaters 92
Hong Kong: Mai Po marshes 75–6, 109
horizontal arches 11
Hormosira 98–9
hummingbirds 90, 91–2
hurricanes and typhoons 163, 204, 207, 210, 220–1, 228
Hydnophytum formicarium 79
Hydrocharitaceae 4–5, 46, 200
Hydrocharitoideae 5
hydroids 100
Hydrophidae (sea snakes) 85
hydrophilous pollination 46
hydrozoa 132
Hymenoptera 198
  see also ants; bees
hyperosmoregulation 123
hypersalinity 21, 42, 55
Hypochrysops (butterfly) 79
hyposalinity 42

Ilyoplax 120
impacts of human activities 206–37
  economic value of mangroves and seagrasses 233–6
  future of mangroves and seagrasses 236–7
  global climate change 229–33
  hurricanes and typhoons 220–1
  Indus Delta 222–8
  pollution and mangroves 216–20
  rehabilitation of mangroves 221–2
  seagrasses 228–9
  shrimps and mangroves 213–16
  sustainable management: Matang (Malaysia) 211–13
  uses of mangroves 208–11
  see also aquaculture; exploitation; fisheries; pollution
India 127, 197
  see also see Sundarbans
Indian Ocean 52, 100, 102, 104, 231
indirect uses of mangroves 209–10
Indo-Malesia 186–8
Indo-Pacific 7, 35, 54, 209
  comparisons and connections between mangroves and seagrasses 168, 178, 182
  marine components 112, 127–8, 134
  seagrass communities 143, 144
  terrestrial components 78, 86
  see also Indo-West Pacific region
Indo-West Pacific region 68, 125, 185–8, 191–4, 203
Indonesia 75, 140, 215, 235
Indus Delta (Pakistan) 9, 17, 24, 207, 222–8, 237
  community structure and dynamics 49–51
  terrestrial components of mangroves 91, 94

inland mangroves 51
insects 73–82
    ants 77–9
    fireflies, synchronously flashing 80–2
    herbivory 73–7
    mosquitoes 80
    termites 77
    see also beetles; bugs
interactions between species 169
    see also competition; succession
intertidal species 2, 4, 39, 57, 67, 145
inundation regime 26
invertebrates 101–30
    molluscs 127–30
    see also crustaceaion
    uptake 21–2
    see also nitrates; phosphates; salt
Iridomyrmex (ant) 79
iron 9
Isoodon spp. (bandicoots) 94
isopods 102, 138, 140

Japan 145, 200
jellyfish 210

K-selection 62
Kandelia 2, 29, 36, 188
    candel 27, 109
Kenya 132, 133, 176, 177–8
kingfishers 89, 118
kinorhynchs 131–2
knee roots 14
Krebs (tricarboxylic acid) cycle 16
Kuwait 222

Labyrinthula zostera (slime mould) 228–9
lactate 16
lacunae 42
Laguncularia 2, 20, 37, 77, 187, 221
    racemosa 25, 36, 113, 196, 220
lampyrid beetles 75
land crab see *Cardisoma*; *Gecarcinidae*; *Gecarcoidea*; *Ucides*
larval dispersal and return 174–5
Lates calcarifer (barramundi) 182
Laticauda colubrina (sea snake) 85
latitudes 153
leaf 19, 21, 23, 74–5
    area 24
    burial 110
    composition see carbon/nitrogen ratio; tannin
    damage 73–4
    eating see crabs, leaf eating; herbivory
    litter 66, 112, 154, 157, 170–1
        see also detritus
    temperature 22–3
leafcutters 77, 79
Leguminosae 71

lenticels 12–15
Lepidoptera 75
    see also butterflies; moths
lianes 71
lichens 71
light 41–2, 68, 151
    see also photosynthesis
lignin 143, 172
lignocellulose 158
litter:
    decomposition 232
    processing 111, 203–4
    production 152–4
    see also decomposition; leaf litter
Littoraria (snail) 190
    angulifera 128
    intermedia 128
    pallescens 129
    scabra 128
Littorinidae 190
Liza spp. (mullets) 133
lizards 86, 88
    see also monitor lizards; *Varanus*
local diversity of mangroves 192–6
Lophogobius 160
Loranthaceae (mistletoe) 72
Lumnitzera 2, 25, 56, 91, 187
    littorea 74
lycosids 83
Lytechinus variegatus (purple urchin) 140–1
Lythraceae 2

Macaca sp. (monkey) 95
Macroglossus minimus (long-tongued fruit bat) 96
Macrophthalmus (ocypodid crab) 120
    hirtipes 140
maize 21
Malaysia 71
    Batu caves 97
    biodiversity and biogeography 187
    community structure and dynamics 52–3
    comparisons and connections between mangroves and seagrasses 178, 180, 182
    impacts 207, 234
    marine components of mangroves 98, 110, 112–13, 117, 120, 125, 131, 132
    measuring and modelling of interactions 151, 157–8, 159
    Merbok 50
    Port Klang/Klang Strait 55–6, 179, 181
    Selangor 80–1, 133, 211
    terrestrial components of mangroves 73, 77, 82–3, 85, 91, 92, 96, 106
    western 11, 173
    see also Matang
mammals 118
manatees 144–5

mangrove trees and their environment 2–4, 8–38
    nutrients, inorganic 24–9
    salt, coping with 17–21
    survival, cost of 21–4
    tropicality of mangroves 38
    waterlogged soil, adaptations to 8–17
    see also reproductive adaptations
mannitol 19
marine components of mangrove community 98–136
    algae 98–9
    fauna of mangrove roots 99–101
    fish 133–6
    meiofauna 130–3
    see also invertebrates
Matang (Malaysia) 50, 61, 203, 206, 233
    measuring and modelling of interactions 149, 152, 156
    sustainable management 211–13
maximum sustainable yield 179
measuring and modelling 147–65
    of mangroves 147–63
        assembling model 161–3
        biomass 148–51
        production estimation 151–4
        tree measurement 148
        see also production of mangroves
    of seagrasses 163–5
medicines 209
Mediterranean 6, 7, 48, 139, 142, 143, 179
Megachiroptera 96
meiofauna 130–3, 157
Meliaceae 2
    see also Xylocarpus
Meliphagidae (honeyeaters) 92
Melita 160
Merguia (shrimp) 125
Mesocapromys angelcabrerai (Cabrera's hutia) 94
methane 9, 231
Metopograpsus 102–3
Mexico 145
    Gulf of 140, 144, 145, 220
microalgae 139, 140, 148
microbial breakdown 154–7
Microchiroptera see bats
midges 80
million acre feet (MAF) 223–4
Miocene era 189
molluscs 99, 100, 127–30, 132, 188
    biodiversity and biogeography 190
    impacts 228, 234
    seagrass communities 138, 139
    see also bivalve molluscs; gastropods; snails
Monacanthidae (leatherjackets) 142
monitor lizards 86–7, 213
monkeys 95, 118, 212, 222
Mormopterus loriae (bat) 96
Morula lugubris (snail) 99

mosquitoes 80, 215
moths 82
Mozambique 53
mud 64–7, 119–23
mudskippers 118, 134–6
Murrayella 99
mussels 130, 139, 160
mycorrhizal fungi 25
Mycteria cinerea (milky stork) 213
Myrmecodia 79
myrmecophytes 78–9
Myrsinaceae 2
    see also Aegiceras
Myrtaceae 2
    see also Osbornia

Nasalis larvatus (proboscis monkey) 95
Nasutitermes 77
Nectariniidae (sunbirds) 92
nematodes 131–2, 133
Neosarmatium:
    meinerti 105, 109
    smithi 104, 107–8
Nephila clavipes (golden silk spider) 83
Nerodia fasciata (snake) 86
New Caledonia 185
New Zealand 7, 140, 145, 221
Nicobar island 75
Nigeria 125
nitrates 9, 24–7, 67, 89, 154, 169, 171
nitrification 25, 26
nitrites 171
nitrogen 25–8, 69
    atmospheric 98
    comparisons and connections between mangroves and seagrasses 168, 169, 171
    -fixing bacteria 99
    gaseous 9, 26, 28–9
    impacts 215, 217, 230
    inorganic 24–5
    marine components of mangroves 100
    measuring and modelling of interactions 158, 161
    seagrass communities 41, 44–5, 143
    terrestrial components of mangroves 79
    see also ammonia; carbon/nitrogen ratio; nitrates; nitrites
nitrogenase enzymes 25
nitrous oxide 26, 29
non-use values 235
Nophopterix (pyralid moth) 75–6
North America 91, 168
    see also United States
nutrients 69, 163, 164, 168, 169
    inorganic 24–9
        and growth 27–9
        nutrient recycling 26–7
        see also phosphorus; nitrogen
    and seagrasses 44–6

Nycticeius greyi (bat) 96
Nyctophilus arnhemensis (bat) 96
Nypa 2, 29, 59, 189, 190–2
  fruticans 187, 189, 209

ocypodid crabs 112, 124, 167
Ocypodidae 102, 104, 113–14, 120
  see also *Ilyoplax*; *Macrophthalmus*; *Uca*;
      Ucides
Odocoileus (key deer) 94
Oecophylla smaragdina (weaver/tailor ant) 78
oligochaetes 132, 133
omnivores 141
Ophiophagus hannah (king cobra) 85
Ophiusa (noctuid moth) 75
Orcaella brevirostris (dolphin) 93
orchids 71
organic particles 159–60
origins of mangroves 188–92
Osbornia 2, 20
osmoconformer 123
osmoregulation 123
osmotic concentration 124
osmotic problems 122–3
ospreys 89
ostracods 140
otters 93, 213
outfalls 225
outwelling 50, 170–1, 178, 180
overgrazing 225, 228
overwash mangroves 51
oxidative phosphorylation 16
oxygen 11–15, 17, 24, 66, 120
  levels, low 8–9, 119–21
  seagrasses 41–2, 48
  see also anaerobic conditions
oysters 65, 130, 160, 222

Pacific 4, 36, 51, 102
  biodiversity and biogeography 200, 203
  east 6, 7, 100, 185–7
  impacts 231, 233
  west 6, 7, 81, 100, 185
  see also Atlantic-Caribbean East
      Pacific region; Indo-Pacific
Pakistan see Indus delta
Palaemonetes (shrimp) 160
Palaeocene era 189
Palaeowetherellia 189–90
Palmae 2
  see also *Nypa*
Panama 54, 89–90, 91, 125, 128, 187, 221, 222
Panthera tigris (Bengal tiger) 94
Papua New Guinea 72, 81, 92–3, 128, 189, 203
  Fly River 50, 64
Parasesarma:
  erythrodactyla 109
  leptosoma 113–14

  maipoensis 109
  plicata 104
Pardosa (lycosid) 83
parental investment 33–4
parrot-fish 143
particulate matter 169, 176–7
Parus major (great tit) 92
passerines 89, 92
patch dynamics 61
pelicans 89
Pelliciera 2, 36–7, 189, 190
  rhizophorae 36
Pellicieraceae 2
  see also *Pelliciera*
Pemphis 2
penaeid shrimps 102, 125, 179, 180–1
Penaeidae 178–9
  see also aquaculture; shrimps
Penaeus 214
  merguiensis 179
  semisulcatus 179
Perameles spp. (bandicoots) 93–4
Periophthalmodon (mudskipper) 135–6
Periophthalmus (mudskipper) 135
  cantonensis 135–6
  sobrinus 136
Perisesarma:
  bidens 109
  eumolpe 105
  onychophora 105, 158
  plicata 105
Persian Gulf 144
pesticides 217
Philippines 133, 140–1, 189, 207, 215, 222, 235
Pholadidae (piddocks) 99
phosphates 24, 27–9, 67, 89, 154, 169
  ferric 26
  ferrous 26
  inorganic 9
phosphorus 26–9, 69, 161, 171, 215, 217
  seagrasses 44–5
  see also phosphates
photosynthesis 2, 12, 21–2, 24, 68, 148, 229–31
  of seagrasses 41–2, 138
Phyllospadix 4, 5, 7, 39, 67, 200
  iwatensis 201
  japonicus 201
  scouleri 201
  serrulatus 201
  torreyi 45
physical barriers 185
physical gradients 55–8, 59
piddocks 99
pigs 94
Pinna nobilis (giant fan mussel) 139
Pipistrellus tenuis (bat) 96
planktonic larvae 118, 169, 175, 181
plants and mangroves 71–2

INDEX

Platanista gangetica (dolphin) 93
platinum electrode probe 8
Plumbaginaceae 2
pneumatophores 14–16, 24, 65, 99
pollination 29–30, 196–7
   hydrophilous 46
pollution 207, 216–20
   oil 218–20
   sewage 217
   thermal 217
polychaetes 100, 132
Polyrachis sokolova 79
porpoises 93
Portunidae (swimming crabs) 102, 124
   see also Scylla
Posidonia 5, 7, 41, 48, 70, 139, 140
   cretacea 200
   oceanica 48
Posidoniaceae 5
Posidonoideae 5
Potamogetonaceae 4–5, 46, 200
potassium ions 42–3, 84, 86
prawns *see* shrimps
predators 139, 140, 142, 160–1
Presbytis spp. (monkeys) 95
primary production 66, 148, 168, 169–71, 174, 178
   autotrophic 147
   gross 152
   net 152, 153, 157, 158, 162, 164
   potential 151
Principe 194–5
Procyon cancrivorus (raccoon) 94
production of mangroves 151–61
   crabs and snails 157–8
   microbial breakdown 154–7
   organic particles 159–60
   predators 160–1
   sediment bacteria 159
   wood 158–9
productivity 203
prokaryotes 98
   see also bacteria; cyanobacteria
proline 19, 44
prop roots 15–16
propagules 30–8, 47–8, 54–5, 57, 76, 222
protein 111
Pseudoapocrytes (mudskipper) 135
Pteridaceae 2
   see also Acrostichum
Pteroptyx (firefly):
   cribellata 81
   malaccae 81
   tener 81
Pteropus (flying fox) 96
Puerto Rico 133, 221
purification 235
pyruvate 16
pythons 85

r-selection 62
raccoons 118
ramets 40–1
Rana cancrivora (crab-eating frog) 83–5
rays 142
recreation 235
Red Sea 167, 186, 232
   northern 16
   Ras Mohammed 51
redox potential 8–9, 15–16, 28, 42, 68
regional diversity in mangroves 184–8
rehabilitation of mangroves 221–2
replanting programmes 207, 228, 237
reproduction and seagrasses 39, 46–7
reproductive adaptations 29–38
   crabs 118–19
   dispersal 34–6
   fecundity and parental investment 33–4
   pollination 29–30
   propagules 30–3
   vivipary 36–8
reptiles 85–8, 118
respiration 41–2, 152
Rhinoceros sundaicus 94
Rhipidura (fantail) 90
   rufiventris 90
Rhithropanopeus 160
rhizomes 39–41, 46, 48, 69
Rhizophora 2, 9, 10, 11, 13–17, 19–20, 22, 27
   apiculata 56, 61, 74, 212–13
   biodiversity and biogeography 187–8, 189, 190
   community structure and dynamics 51–2
   harrisonii 36, 54, 196
   impacts 219, 221, 222, 227
   mangle 11, 25, 27, 30–2, 36, 186
      biodiversity and biogeography 196
      community structure and dynamics 54, 60
      impacts 217, 220
      marine components of mangroves 99, 113
   marine components of mangroves 99, 100–1, 113, 132
   measuring and modelling of interactions 149–52, 156, 158, 164
   mucronata 31, 55–6, 109, 113, 228
   racemosa 196
   reproductive adaptations 29, 31–8
   samoensis 185
   stylosa 30, 57, 74
   terrestrial components of mangroves 73–6, 78–9, 94–5
   x lamarcki 74
Rhizophoraceae 2, 3, 31, 188
   see also Bruguiera; Ceriops; Kandelia; Rhizophora
Rhodophyta 98
rias 51
righting response time 123

riverine forests 51, 208
rivers 169
rodents 94
root:
   architecture 16
   function 15–16
   /shoot ratio 21
   structure 15–16
   structure of mangroves 10
   see also aerial roots
Rubiaceae 2

sabellid annelid worms 100
salinity 17, 23, 36
   community structure and dynamics 51, 55, 57–8, 62, 66
   comparisons and connections between mangroves and seagrasses 168
   impacts 207, 217, 224, 230
   marine components of mangroves 129
   and seagrasses 39, 42–4, 48
   terrestrial components of mangroves 84–5
   see also hypersalinity; hyposalinity
Salmacis sphaeroides (sea urchin) 141
salt 24
   coping with 17–21
   glands 19, 66, 88
   marshes 168–9
   secretion 42
   tolerance 23, 72
   see also salinity
Samoa 185
sandflies 80
São Tomé 194–5
sap 209
saprophytic fungi 79
Sarpa salpa 142
satellite imagery 224–5
Scaridae (parrot fish) 142, 178
Scartelaos (mudskipper) 135–6
Scarus (parrot fish) 142
   guacamaia 178
scrub mangroves 51
Scylla sp. (mud crab) 124, 127, 210, 215
Scyphiphora 2
   hydrophyllacca 74
sea level 51, 59, 229, 232–3
sea urchins 138, 140–1, 143, 228
seagrasses 4–7, 39–48, 67–70, 137–46
   alteration of environment 48
   birds 145–6
   crustaceans 140
   dugongs and manatees 144–5
   echinoderms 140–1
   epiphytes 137–8
   fish 141–2
   growth and structure 39–41
   impacts 228–9
   molluscs 139
   nutrients 44–6
   photosynthesis and respiration 41–2
   propagule dispersal 47–8
   reproduction 46–7
   salinity 42–4
   turtles 143–4
   see also *Amphibolis, Enhalus, Halodule, Halophila, Phyllospadix, Posidonia, Syringodium, Thalassia, Thalassodendron, Zostera*
seasonal variations 170
secretion 19
sedimentation 64–5, 115, 117, 176
   bacteria 159
   composition 58
   impacts 207, 221, 224, 228, 229, 232, 233
   mangroves 50
   seagrasses 45, 48
seedlings 110–13, 228
   see also propagules
self-pollination 196–7
Senegal 95
serpulid annelid worms 100
Sesarma 102–3, 122
   meinerti 80
   reticulatum 121
sesarmid crabs 168, 187–8
   marine components of mangroves 102–5, 107, 109, 112, 120–1, 124
   measuring and modelling of interactions 15, 157–8, 161
   see also *Aratus; Armases; Cleistocoeloma; Episesarma; Neosarmatium; Parasesarma; Perisesarma; Sesarma; Sesarmoides*
Sesarmidae 102
Sesarmoides kraussi 105
settling velocity 64
shade intolerance 61
sheep 225
shipworms 130, 158–9
shrimps 125, 138, 140, 160, 178–82, 213–16
   fisheries 210, 226
Siganidae (rabbitfish) 142
Sinai Desert (Egypt) 14, 16, 51
Singapore 82, 91
snails 99, 127–30, 157–8
   see also gastropods
snakes 88, 118
sodium ions 42–4, 84, 86, 88
soil:
   composition 26
   oxygenation 26, 68
   pH 58
Sonneratia 2, 14, 19–20, 23, 29
   alba 19, 23, 96, 109
   caseolaris 81, 96
   community structure and dynamics 56, 59
   lanceolata 19, 23
   marine components of mangroves 132

Sonneratia (Cont.)
  measuring and modelling of interactions 156
  ovata 96
  terrestrial components of mangroves 95–7
Sonneratiaceae 2
South Africa 105, 153
South America 4, 50, 77, 94, 189
  see also Amazon
south-east Asia 31, 62, 80, 83, 85–7, 90–2
  biodiversity and biogeography 188, 189, 200, 201
  impacts 207, 209
  marine components of mangroves 118, 125
  measuring and modelling of interactions 150
  seagrass communities 144
  terrestrial components of mangroves 80, 83, 85–7, 90, 91, 92, 93, 95
Sparidae 142
Sparisoma spp. 142
Spartina 126, 168
species:
  -area relationships 184, 194, 198
  diversity 145, 147–8, 184
  richness 203–4
  zonation 52–61, 63, 99, 119, 232
    geomorphological change 59
    physical gradients 55–8
    plant succession 59–61
    propagule sorting 54–5
Sphaeroma 99–100
spiders 82–3
sponges 25, 100
Squilla choprai 125
Sri Lanka 182, 189
stable isotope ratio 17, 106–7, 128, 130, 172–4, 176–8, 180–1
starch 37
stenohaline species 57
Sterculiaceae 2
  see also Heritiera
stilt roots *see* aerial roots
stochastic factors 194
stomatal conductance 22
Stomatopoda (mantis shrimps) 124–5
storks 89
stress, environmental *see* anaerobic conditions; desiccation; salinity; temperature
Strombus gigas (queen conch) 139
structure of seagrasses 39–41
subtidal species 2, 39, 42, 68
succession 59–61, 62–4
sulphate 9
sulphide 9, 16, 126
sunbirds 91, 92
Sundarbans 49, 59, 93, 94, 161, 208
Surinam 92

survival, cost of 21–4
sustainable management 206
  see also Matang (Malaysia)
swans 145
swimming crabs 102
Syngnathidae 142
Syringodium 5, 7, 47, 142, 200
  filiforme 43–4, 201
  isoetifolium 201

Taiwan 216
tannins 26, 74, 107–11, 114, 132, 156, 174, 209
Tanzania 53
Taphozous (bat):
  flaviventris 96
  georgianus 96
tardigrades 131, 132
tectonic movements 59, 191–2, 200
Tectura depicta (limpet) 139
Tedania (sponge) 25
Telescopium (snail) 127, 209
temperature 122, 184–5, 221, 229, 230
  see also global warming
Terebralia (snail) 127, 133
  palustris 127–9
teredinid molluscs 158
Teredinidae (shipworms) 99, 130
termites 77
terrestrial components of mangrove community 71–97
  plants 71–2
  see also animals from the land
thaidid molluscs 100
Thailand 81, 119, 144, 152, 207, 216, 222, 235
Thais (snail) 127
  kiosquiformis 99
Thalassia 5, 7, 46, 48, 176, 200
  hemprichii 47, 68, 141, 144, 201
  testudinum 40, 43, 44, 69, 143, 164, 201
Thalassina (mud lobster) 61, 102, 125–6
Thalassinids 140
Thalassioideae 5
Thalassocharis 200
Thalassodendron 5, 7, 47, 176, 200
threats, indirect and accidental 207
tides/tidal 169, 218
  -dominated estuaries 50–1
  fluctuations 67, 177, 193–4
  -inundation regime 55–6
tissue tolerance 42
tolerance 19
tortricid moth 78
tourism 213
  see also ecotourism
transpiration 42
Trichecus (manatee):
  manatus 144
  senegalensis 144

Trichoderma 157
Trinidad 92, 235
   Caroni Swamp 89, 211
Tripneustes gratilla (sea urchin) 141
Trochilidae (hummingbirds) 92
tropicality of mangroves 38
tsunamis 210, 220
tunicates 100
turbellarian flatworms 130–1, 132
turtles 138, 143–4
typhoons *see* hurricanes and typhoons

Uca (fiddler crabs) 102, 105–6, 114–20, 124, 126, 133, 160, 168–9, 175
   dussumieri 116, 117
   lactea 116–17, 119, 122, 128
   lactea annulipes 117
   polita 117
   pugnax 117
   rosea 118
   tangeri 95
   urvillei 116
   vocans 117
Ucides (land crab) 104, 113
ultraviolet radiation 67
Ulva (algae) 217
understorey vegetation, lack of 62–3
United Kingdom 190
United States 222
   Bodega Bay (California) 205
   California 145, 179, 187
   Maryland 190
   Monterey Bay (California) 139
   see also Florida
urea 85–6
use values, direct and indirect 235
uses of mangroves 208–11

Vanda (orchid) 71
Varanus indicus (monitor lizard) 87
vascular plants 71
Venezuela 101
vertebrates 83–97
   amphibians 83–5
   birds 89–93
   mammals 93–7
   reptiles 85–8
vertical sequence of species 52

Vietnam 207, 215, 220
vivipary 30–1, 33, 36–8, 47–8

Wadden Sea (Holland) 145
waders 89
warblers 90, 91
water:
   buffalo 225
   economy 122
   expenditure minimization 23
   uptake 21–2
   use, changes in 207
   -use efficiency 22
waterfowl 145, 146
waterlogged soil 8–17, 24
   see also anaerobic conditions
wave:
   action 50, 228
   conditions 218
   -dominated estuaries 50
West Indies 51
Wetherellia 189
wind conditions 218
wood 158–9
   -borers 100
woodpeckers 90

Xanthidae (crabs) 112
Xeromys myoides (rat) 94
Xylocarpus 2, 14, 19–20, 29, 31, 33, 231
   australensis 74
   granatum 10, 14, 74
   mekongensis 10

yeast 16
Yemen 51

Zostera 4–5, 7, 44, 47, 67, 139–40, 145, 200
   asiatica 41, 201
   capricorni 45, 67, 145
   caulescens 201
   japonica 201
   marina 6, 39–42, 47, 70, 141, 145, 201, 205, 228
   noltii 145
Zosteraceae 5
Zosteroidea 5

QK
938
.M27
H64
2007